The Kandik Map

The Kandik Map

Linda Johnson

University of Alaska Press
Fairbanks, Alaska

University of Alaska Press
PO Box 756240
Fairbanks, AK 99775-6240

ISBN 978-1-60223-032-3 (cloth)
ISBN 978-1-60223-042-2 (paper)

Library of Congress Cataloging-in-Publication Data

Johnson, Linda, 1949 May 21–
The Kandik map / by Linda Johnson.
p. cm.
ISBN 978-1-60223-032-3 (cloth : alk. paper)—ISBN 978-1-60223-042-2 (pbk. : alk. paper)
1. Yukon—Maps. 2. Cartography—Yukon—History—19th century. 3. Kandik, Paul, fl. 1880. 4. Mercier, François, fl. 1868–1885. 5. Fur trade—Yukon—History—19th century. I. Title.
GA475.Y8J64 2008
912.798'6—dc22

2008018450

Cover and text design by Paula Elmes, ImageCraft Publications & Design

Cover Art: Alfred Boisseau (1823–1901), *Explorer in the Arctic*, 1887, Oil on canvas, 25 x 41 inches. Collection of Morris Communications Co., Augusta, Georgia

This publication was printed on acid-free paper that meets the minimum requirements for ANSI / NISO Z39.48–1992 (R2002) (Permanence of Paper for Printed Library Materials).

Contents

Acknowledgments

MANY PEOPLE have contributed to the realization of this book. I was inspired to look for documentary sources and related oral traditions about early encounters between northern indigenous people and newcomers by my earliest mentors in the Yukon. These include First Nation Elders and pioneer researchers who visited the Yukon Archives on a regular basis when I first arrived to work there in 1974, and many more who enriched my understanding of northern people and landscapes throughout my subsequent career as an archivist and researcher. There is room here to acknowledge only those who contributed most directly to my Kandik Map research, but I wish to extend warm thanks to all those who have so generously shared information and ideas through the years. Some have passed on in recent years and they are greatly missed.

Among the Elders, I thank Angela Sidney, Annie Ned, Austin Hammond, Effie Kokrine, Elijah Smith, Frances Woolsey, George Dawson, Isaac Juneby, Julia Morberg, Kitty Smith, Percy Henry, Violet Storer, and many more. Fellow researchers include Adeline Peter Raboff, Allen Wright, André Mercier, Andy Bassich, Bessie and Bonar Cooley, Catharine McClellan, Craig Mishler, David

Neufeld, Frederica de Laguna, Georgette McLeod, Gerald Isaac, Jim Kari, John Ritter, Julie Cruikshank, Lani Hotch, Lori Eastmure, Louise Profeit Leblanc, Lulla Johns, Madeline de Repentigny, Mike Gates, Natascha Sontag, Norm Easton, Phyllis Fast, Syri Tuttle, Willem de Reuse, Yann Herry, plus many other colleagues. I acknowledge the contributions of the University of Alaska Fairbanks Northern Studies Program—Judy Kleinfeld, Mary Erhlander, Julia Parzick, faculty and students—and my Graduate Advisory Committee who guided my thesis on the Kandik Map—William Schneider, Mary Mangusso, and Phyllis Morrow.

I extend my deepest appreciation to all the staff at research and cultural institutions which preserve and make available essential northern sources, including the Alaska & Polar Regions Collections at the University of Alaska Fairbanks, the Alaska Native Language Center, l'Association franco-yukonnaise, the Bancroft Library, Eagle Historical Society, Eagle Village, Special Collections Department at Gonzaga University, the Hudson's Bay Company Archives at the Manitoba Archives, Library and Archives Canada, Tr'ondëk Hwëch'in First Nation Heritage Department and Dänojà Zho Cultural Centre, Yukon Archives, Yukon College Library, and the Yukon Native Language Centre.

Finally, I thank my family for steadfast support through many years of travel, study, and writing: my husband David, sons Galen and Daniel, mothers Wanita and Doris, sister and brother-in-law Jane and Gary.

Thank you, *merci beaucoup, massi cho!*

Linda Johnson
Whitehorse, Yukon
March 2009

1

The Kandik Map: Reflections on Time and Space

TRAVELERS TODAY drive through the winter snows or summer splendor of the Alaska-Yukon borderlands on the Alaska Highway, going from one warm, well-lit community to the next in just a few hours, with the certainty of groceries, gas stations, motels, emergency care, and year-round residents to offer survival, communication, and comfort on a continuing basis. In summer, the more adventurous can hike the Fortymile uplands, fly into Yukon-Charley Rivers National Preserve to kayak on a wild creek, or cruise from Dawson to Eagle on a high-tech catamaran, while winter explorers can follow the Yukon Quest Trail by snowmobile or dog team along the frozen Yukon River. Both longtime residents and newly arrived visitors may marvel at the services available to travelers in this vast land, asking who built these roads and communities, who passed this way before, and who lives here now and why. Few travelers, whether Native or nonnative resident of Alaska or Yukon, or recent visitor, will know or learn that this is the Kandik Map[1] country—the land where an Indian man called Paul Kandik and a French Canadian named François Mercier lived and traveled more than a hundred years ago, combining their knowledge in 1880 to create the earliest known map of the area. Why have their story and their contributions to the cartographic knowledge of this region faded into obscurity? What can we learn today about their lives, their map, and the significance of their contributions? Where and how can their stories be found? What does it mean to revive a document—to bring new life to the stories surrounding it? Who will listen and what will they learn?

Discovering the North: Story Lines and Camps

"Discovering" and "building" the modern North have long been central themes in popular and academic writing about Alaska and the Yukon, with stories of the heroic deeds of past generations of explorers and settlers commemorated in books, songs, plays, television and radio programs, monuments, road signs, maps, and visitor brochures. Most of this historical reflection has focused on the lives and work of nonnative newcomers, celebrating the first white man to cross the Chilkoot Pass, the first white woman to winter in the Fortymile, the first white baby born in the North, and other "first" passages through river valleys, mountain passes, seasons, and epochs. These "first northerners" are represented as true pioneers, enduring brutal extremes of cold, hunger, mosquitoes, isolation, and sometimes hostile Native peoples, to discover the North and transform its wild spaces from "terra incognita" into "known" places and "civilized" communities marked on maps and described in reports.[2] Often the exploits recorded about and information gathered by these "first" newcomers were possible only with assistance from resident Native guides for whom this knowledge was a matter of everyday wisdom passed down from countless generations of ancestors. The names of those Native guides were seldom recorded, their contributions often unacknowledged by the newcomers whom they assisted, sometimes because of communication and translation difficulties which rendered their Native names "unpronounceable," sometimes owing to ethnocentric attitudes common among newcomers who dismissed their guides' concerns regarding the dangers of routes proposed or season of travel, and often because explorers were hungry to claim the "discoverer's" right to fame and publicity.[3] The information transmitted in records and sketches by early explorers and officials was assumed to be correct by virtue of their being reputable "authorities"—usually the agents of national governments, either military personnel or civilian surveyors, or of religious and commercial organizations allied in the cause of discovering and exploiting the resources of the North.[4] In turn those authoritative claims to discovery and knowledge have informed the opinions of several successive generations of writers and scholars, building up an enormous body of information about northern lands and peoples, presumed to be accurate by the sheer weight of accumulated time, talent, and evidence.[5]

The stories told through northern highway signs and maps, visitor brochures, museum exhibits, walking tours, and community residents vary from place to place, and may differ from those of "outside" specialists and scholars. Most of these stories lead down certain well-blazed trails, repeating facts, ideas, themes, and conclusions long embedded in oral traditions, archival records, and literature, but often separated in distinctive story "camps," based on the different linguistic, cultural, and memory traditions of various resident northerners and "outside" commentators.[6] These story "camps" reflect debates within and outside of Alaska and the Yukon about the origins, history, ownership, purpose, meanings, and development of the region, topics which have been the focus of both community discussions and scholarly discourse for a very long time.

Recovering the North: New Sources and More Perspectives

In recent decades, the variety, sources, and audiences for these stories have been expanding, with new recording, duplication, and broadcasting technologies, plus political and scholarly developments, supporting the emergence and documentation of oral traditions and community histories to clarify, enlarge, and enrich the concepts of northern life.[7] Some key events and subsequent shifts in perception have assisted this process—notably the deliberate decisions and ongoing efforts of aboriginal language speakers, tradition bearers, and indigenous organizations throughout Alaska and the Yukon to document and share their knowledge with a broad public beyond their original communities.[8] In storytelling festivals, land claims negotiations, pipeline hearings, climate change testimonials, publications, radio and television broadcasts, films, and a host of other venues, Elders and other Native people communicate their perspectives about issues affecting their lands, their lives, their cultures and histories, and their dreams for future generations.[9] Native authors are bringing important family and community stories to public notice.[10] Books such as Adeline Peter Raboff's *Inuksuk*[11] masterfully combine oral traditions and documentary sources to tell a powerful history, with linguistic and cultural contexts rarely available to people outside Native communities.

Nonnative commentators have provided significant intellectual and contextual support for these efforts as well, contributing to a new public receptivity for the stories of indigenous people both in the North and in the world beyond—a legitimacy that is long overdue and difficult to achieve in the face of overwhelming historical and socioeconomic pressures. Canadian Justice Thomas Berger's reports, *Northern Frontier, Northern Homeland*[12] and *Village Journey*,[13] delineated many of the factors contributing to the historical and conceptual gulf between indigenous and newcomer perspectives that frustrated comprehensive community approaches to northern issues for many years. He provided a social justice and scientific policy rationale for considering indigenous knowledge for the benefit of all northerners and outsiders with an interest in the North. Julie Cruikshank worked with Yukon Athabaskan women to record and publish their stories in *Life Lived Like a Story*,[14] offering a new lens for viewing events such as the Klondike Gold Rush and the building of the Alaska Highway as experienced and interpreted by these women. Other community histories also contribute to a more nuanced and multicultural understanding of northern history—l'Association franco-yukonnaise has published several books documenting the lives of francophones in the North,[15] while Alaskan family histories such as Judy Ferguson's *Parallel Destinies*[16] give voice to settlers of east European origins and their integration within their new homeland. Numerous local history projects in Alaska and the Yukon offer similar insider stories and culturally constructed northern viewpoints that go beyond, and often challenge or even contradict, the perennial versions of historical events, their consequences and meanings.[17]

In *The Social Life of Stories*,[18] Cruikshank reexamined several Yukon historical episodes that form the foundation for much of the popular and scholarly writing about the Klondike Gold Rush, comparing the two separate story lines handed down in Native and nonnative traditions. By recording and publishing the stories of the Tagish man Keish with his niece Angela Sidney, Cruikshank provided details of his background and motivations as George Washington Carmack's partner on Bonanza Creek that contribute many new meanings to the events of 1896. He was no longer simply Skookum Jim as portrayed at the time by William Ogilvie[19], who questioned him closely to assess the truthfulness of his claim to the original discovery of gold, and later writers like Pierre Berton,[20] who repeated Ogilvie's version and added many more accounts from nonnative stampeders. Sidney's history tells about Keish's family concerns for a missing sister which caused him to make the long trip from Tagish to the Klondike, his spiritual beliefs and practices which led him to rescue a frog who later appeared to him in a dream to guide his discovery of gold, and his care for his family and clan after the Gold Rush that motivated his spending of the wealth accrued from his role in starting the "greatest gold rush in history." It is an altogether different window on the times, a departure from the oft-repeated tales, and a fascinating entrée into the world of Tagish people. It is an essential component of the event that changed the Yukon forever, but it was inaccessible to all but Tagish speakers and a few others in the Yukon First Nations community who could understand and appreciate this other view of Keish and his actions. Over time it became inaccessible to younger generations even in that community, as English replaced the Tagish language, and traditional beliefs were discounted and discouraged by missionaries, schools, and the new dominant society in the North. For decades Keish and his Frog Helper story were locked in the linguistic and cultural solitude of his traditional community, separated from mainstream historical interpretation, only vaguely acknowledged in nonnative accounts of the Gold Rush. After the story was documented and readily available,[21] many accounts persisted in retelling the story of "Who found the Klondike gold?" as the Carmack-Henderson saga of a century ago.[22]

A similar focus on nonnative "discovery" stories predominates for the borderlands region of Alaska-Yukon, with numerous reports from the nineteenth century and subsequent twentieth-century history books, detailing the exploits of Hudson's Bay Company men like Robert Campbell and Alexander Hunter Murray, the dedicated sacrifices of missionaries like Archdeacon Robert McDonald and Bishop Bompas[23] traveling in the wilds to bring the word of God to Native people, the daring and dangerous river and overland explorations of Lieutenant Frederick Schwatka and Lieutenant Henry Allen,[24] and the tough prospecting voyages of Jack

McQuesten, Arthur Harper, Al Mayo, and Joe Ladue.[25] Morgan Sherwood's book *Exploration of Alaska,* written in the 1960s, illustrates the progression from the published reports and archival records of nineteenth-century explorers to later historians' writing about the events. In turn both the period and later retelling of those stories informed public perceptions of those times and places, leading to the naming of highways and other commemorations that further reinforced the idea of nonnative "discovery" of the region.

Sherwood's retelling of the "discovery" of the Tanana River is a case in point. In the chapter entitled "The Remarkable Journey of Henry Allen"[26] he praised Allen's prowess in surmounting incredible odds on several fronts—first in dealing shrewdly with local Native people to hire them as packers and guides, and then as "first discoverer" for his trip up the Copper, over to the Tanana, and down to the Yukon. Sherwood celebrated Allen's victories in completing "an original exploration of about 1,500 miles, an exploration that crossed the Alaska Range . . . [charting] three major river systems . . . for the first time." He quoted Mendenhall: "No one geographer in recent years has made greater contributions to our knowledge of the Territory in so limited a time in the face of such obstacles."[27] Sherwood concluded that Allen's voyage "ranks with the earlier investigations of Alexander Mackenzie and Robert Campbell in the Far Northwest and was certainly the most spectacular individual achievement in the history of Alaskan inland exploration."[28] Allen was guided by the knowledge and advice of Native people he met throughout the region, and was saved from starvation and disaster on several occasions by Native women and men who offered his party food and lodging. Allen himself reported their assistance, for the most part in grateful recognition of the essential help offered by them, shaded with some typical nineteenth-century grumblings about the prices charged for food and services in the northern wilderness.[29] Later commentators overlooked the Native helpers he named or relegated them to minor inconsequential roles.[30]

The combined effect of nineteenth-century and later historical writing on northern communities has been profound in the past, leading to imbalance in commemorative naming and interpretation[31] and school curricula that overlooked Native contributions, as well as more serious political battles over land ownership and resource allocations. Recent linguistic documentation and other projects conducted by Native organizations and others in Alaska and the Yukon have contributed new data and perspectives based on oral traditions.[32] The process of recording oral traditions to enlarge the available historical narrative is complicated by modern social conditions in indigenous communities. Rapid, profound, and pervasive socioeconomic changes have accumulated over the past century—resulting in the diaspora of younger people from traditional villages to disparate urban settings, separations between constituent groups of traditional linguistic and cultural communities with the demarcation of international borderlines, and the imposition of new belief systems and values by church and state authorities.[33]

The Social Life of a Northern Document

All of these themes are significant factors in the unveiling of the Kandik Map and in discovering more information about the lives of its creators—the "social life" of the document they created. The two distinctive story "camps" figure in this documentary saga. Paul Kandik's identity and legacy reside solely in the one map bearing his name, supported by oral traditions for the northern Athabaskan community as a whole, but without any additional extant records specific to him. François Mercier's legacy is steadily growing with more documentary and oral sources emerging through linkages between the modern francophone community in the Yukon and his original home in Québec, aided by modern information technologies that make it possible to "Google" Mercier across vast distances, between different linguistic traditions, and beyond national boundaries.[34] Mercier shares a heritage with Kandik in being part of a minority group in the North, soon overtaken and eventually pushed out of the fur-trading business he so proudly represented for over sixteen years in Alaska and the Yukon. His experience in Alaska-Yukon is a mirror to that of French Canadian history in the American West where place-names like Coeur d'Alene linger as artifacts, their origins lost in translation to successive generations of American settlers with no linguistic affinity to the names, nor means to access information about the people they represent from earlier times.[35]

The information Kandik and Mercier recorded was transported to distant places, stored and forgotten for many decades, so that now it is scarcely recognizable by present generations in either northern Native or

Québec communities. The map at the Bancroft Library is an archival orphan today—isolated from context, separated from both of its original linguistic and cultural traditions. Peeling back the layers of obscurity surrounding its creation, meaning, and purpose has revealed extensive information about Mercier; we hope the same will hold true for Kandik over time. Meanwhile modern information technology fosters wide circulation and discussion of the map, its creators, and its possibilities for regenerating community knowledge, pride, and continuity, plus new connections between the descendents of the two cultures that created it, as well as all travelers in the region in years to come.[36] Perhaps the names of Kandik and Mercier will be added to highway signs, interpretive talks, and brochures in future—so that they receive credit for the trails they blazed for others who came later. The Kandik Map has potential for "recovering" the North—for both those whose homeland it is and always was, and those for whom it was a frontier and now is home.

The Kandik Map drawn in 1880 by Yukon Indian Paul Kandik and annotated by French Canadian fur trader François Mercier is a unique record in the mapping history of Alaska and the Yukon. Tracing the Yukon River from above Fort Selkirk downriver to Anvik for some two thousand miles, Paul Kandik's drawing includes the major tributaries and other landscape features en route, as well as the Tanana River and the upper reaches of the Kuskokwim River. The lettering on the map, attributed to François Mercier, includes information in a number of languages—Hän, Tanana, Northern Tutchone, French, and English—documenting placenames, and details of travel times, plus food resources. Some symbolic figures identifying trading posts and other features may have been drawn by Kandik or Mercier, or perhaps by some other traders or unidentified persons. There are annotations by Ivan Petroff, a census taker for the United States Government in 1880 and later one of the authors of Bancroft's *History of Alaska* published in 1886.[37] There are also more recent annotations by the staff of the Bancroft Library, where the original document is located at the University of California, Berkeley.[38] Numerous historical, ethnohistorical, and ethnographic sources document the lives and times of the people living in the region when this map was drawn.[39] Other sources trace the development of geographic and cartographic knowledge about northwestern North America.[40] All of these pieces of evidence contribute to understanding the origins of the map, but are not enough in themselves to tell its story. Local and traditional knowledge of both Native and nonnative people living in the region today is also vital to understanding the history, significance, and meanings of the Kandik Map.

The map is possibly the first recorded drawing of the upper Tanana and Kuskokwim rivers, together with the trails used by aboriginal people to travel between those rivers and the Yukon.[41] It was an important contribution by a Native man to the contemporary nonnative geographical knowledge of this region. While Kandik's background, experience, and knowledge were worlds apart from that of his nonnative colleagues, his reasons for drawing the map may have merged with their economic interests during this time of rapid change. In many ways, this intriguing northern document is reflective of the dynamic era in which it was drawn, serving

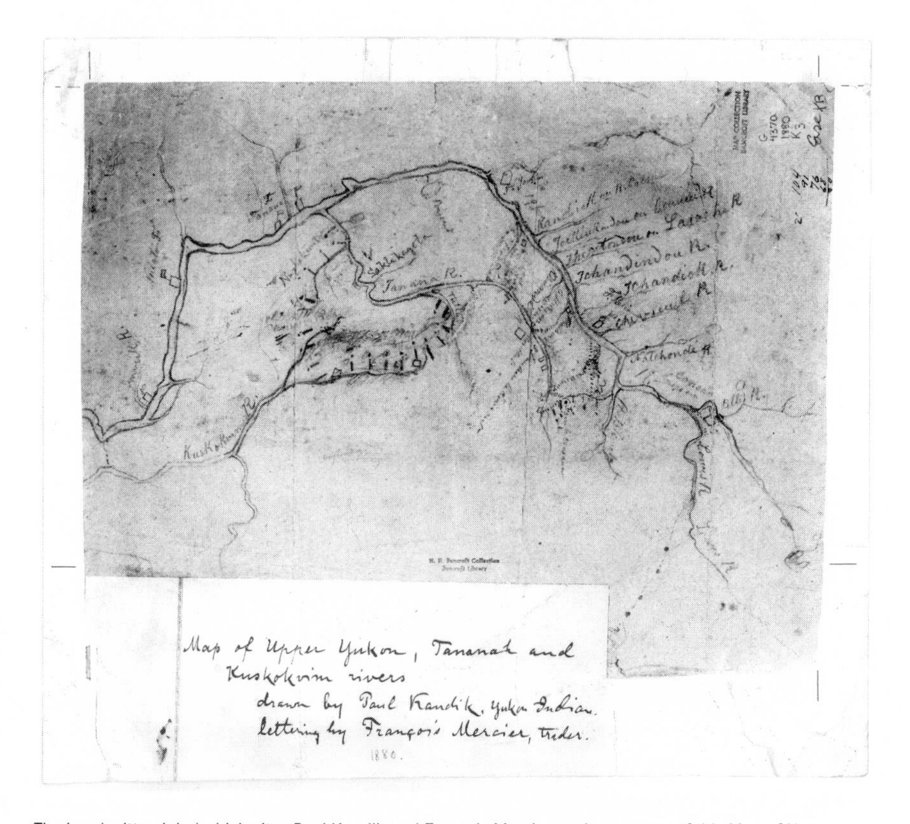

The handwritten label which cites Paul Kandik and François Mercier as the creators of this Map of Upper Yukon, Tananah and Kuskokwim rivers 1880 is the only source that names Paul Kandik and identifies him as a "Yukon Indian." The Bancroft Library, University of California, Berkeley, Map Collection, G4370 1880 K3 Case XB.

as an archival marker or snapshot of the evolving new society in the area. Delving into the history of its creation, travels, provenance, preservation, and significance illuminates both the early days of Native and nonnative interactions in the border regions of Alaska-Yukon and subsequent developments to the present.

The map was drawn at a time when the northwestern part of North America was in a state of flux, just a half-dozen years before gold discoveries at Fortymile would set the stage for the Klondike Gold Rush at the end of the nineteenth century. Athabaskan people had lived along the Yukon River and its tributaries for thousands of years and countless generations, supported by far-ranging aboriginal trade networks and technologies; developing their own social, cultural, and belief systems; maintaining an intensive knowledge of local landscapes and resources; and communicating in distinctive aboriginal languages about concepts of land use, boundaries, meetings, travel, and exchange.[42] European exploration of the Pacific Northwest began with the voyage of Vitus Bering in 1741 on behalf of the Czar, extending Russian territorial interests into the unknown lands beyond the eastern bounds of the empire, setting the stage for the exploits of Russian fur traders in future years. Bering caught sight of a huge mountain along the coast, which he named Saint Elias. It became one of the pivotal landmarks for European and later American and Canadian boundary claims in this region, the point from which the 141st meridian was drawn as a borderline from the Pacific to the Arctic Ocean by distant mapmakers, although these lands would be unknown and untraveled by nonnative people for more than a hundred years after Bering's voyages.[43]

For many decades the Native people of the Alaska-Yukon interior had only indirect contact with European and North American nonnative explorers and traders through coastal aboriginal traders, who extended the influence of the newcomers inland, bringing new tools, clothing, and ideas that shifted social and economic patterns throughout the region. These exchanges gradually set the stage for the later explorations and settlements by nonnative newcomers themselves that would completely transform the lives of interior aboriginal people and the future of their homeland. By the mid-1800s the traders of the Russian-American Company were permanently established all along the coast of Alaska and venturing some way into interior Alaska, while Hudson's Bay Company traders pushed west across British North America to explore the Pelly, Porcupine, and upper

Yukon rivers. American traders competed with both Russian and British traders along the coast, lobbying their government to support further expansion of American influence and interests, culminating in the purchase of Russian Alaska by the United States in 1867, the same year that the new Canadian Confederation was founded. Trading options for interior Native people changed rapidly as new American entrepreneurial ventures and independent traders replaced earlier Hudson's Bay Company and Russian-American Company fur traders. A new line of trade goods came into the country, along with prospectors and miners from the American and Canadian West, lured by reports of gold to be found in the North. Technology transformed river travel as American traders introduced steam-powered boats that traveled much faster than the skin boats and birch-bark canoes of the aboriginal people.[44]

The geopolitical boundary dividing these northwest lands into what is now Alaska and the Yukon was drawn on maps by European diplomats

Tanana River people annually traveled in their birch-bark canoes down to Nuklukayet to trade with Yukon River people and the white traders at the Alaska Commercial Company and other company posts. Here three men stand on the riverbank clothed in a combination of skin garments and trade goods. Photographed c. 1883 by Charles Farciot, one of the earliest photographers to visit the upper Yukon River and inventor of the steam canoe seen to the right of the other canoes. Alaska State Library, Wickersham Collection, Charles Farciot, ASL-P277-017-035.

in 1825, anchored by Bering's beacon, the massive Mount Saint Elias. There were no physical markers on the ground and initially there was no effect on travel, trade, or people along the rivers bisected by the boundary. The northwest interior of Alaska-Yukon was rugged terrain, far distant from the headquarters of the competing fur trade companies. Russian and British traders operated in mutual isolation along the Yukon River for decades, the former ignoring incursions by the Hudson's Bay Company into Russian territory, including the establishment of a permanent trading post at Fort Yukon in 1847.[45]

After the American purchase of Alaska in 1867 the borderline at the 141st meridian took on increased significance. In 1869, United States Army Captain Charles Raymond was sent by the American government to survey the position of Fort Yukon. He traveled up the Yukon with an American fur trader, on a steamboat owned by one of the new American companies interested in pursuing trade in the region. Raymond determined that the Hudson's Bay Company (H.B.C.) traders were located well to the west of the boundary and ordered them to vacate, claiming the buildings and site as property of the United States, which he turned over immediately to the American traders with him. The H.B.C. traders moved up the Porcupine River to Rampart House, from where they tried to compete with the Americans, maintaining contact with their Athabaskan trading partners, hoping through them to control the furs harvested in the region. However, Raymond's longitudinal measurements at Fort Yukon, and subsequent assertion of American claims to the post and surrounding territory, were pivotal events that forever changed economic and social relationships throughout the region. The new American trading companies continued to assert their economic interests by virtue of U.S. territorial claims along the Yukon River, while the H.B.C. enterprise languished on the Porcupine.[46]

François Mercier was among the newcomers arriving on the upper Yukon with the establishment of American control in 1868, accompanied by his brother Moïse. Their French Canadian linguistic and cultural heritage provided important elements of continuity with the previous H.B.C. regime, which had brought both English and French speakers to live and trade among the Native peoples of the region. François had been employed as a fur trader in Montana and Wyoming in the 1860s, giving him experience with the values and methods of trade on the American frontier. By his own definition, he was a fur trader at heart. He stayed in the North for over a decade, struggling fiercely to gain a competitive edge for his sponsor companies. He established new posts and trading partnerships among

François Mercier posing in a classic "coureur de bois" outfit. Illustration from François Mercier, *Recollections*, back cover.

remote Native groups, promoted the expansion of steamboat travel, and was instrumental in recruiting new employees for his trading posts who instituted more significant changes. Ironically, all of these innovations hastened the transition from the old fur trade to a new mining economy along the Yukon, perhaps playing a part in his decision to leave the North in 1885. Mercier's unique background, and his position as an experienced trader along the Yukon, also constituted the perfect combination of interests, knowledge, and contacts for him to play a pivotal role in the creation of the Kandik Map.[47]

For Native people in the region these events signaled more change, at a more rapid pace, especially with technological innovations such as steamboats, repeating rifles, and other American trade goods. The changes also brought new economic opportunities and adjustments, along with the continuing challenges of disease and social upheaval. For many Native people the borderline at the 141st meridian probably continued to be both unfamiliar in concept and unimportant in reality most of the time and in most places. For indigenous people on the upper Yukon and the Tanana who were used to trading with the H.B.C. at Fort Yukon, the American assertion of authority did affect their travel patterns, trade locations, and partners, immediately after Raymond's visit in the summer of 1869 and on a continuing basis thereafter. The arrival of the Americans, with their steamboats, competitive trading practices, and new prospecting focus, undoubtedly influenced the interests, knowledge, and contacts of many Native men like Paul Kandik. He may have been a very young man at the time of Raymond's visit when travel along the whole length of the Yukon was limited by social and technological factors.[48] By 1880, he was able to represent a broad concept of the geography of the Alaska-Yukon interior, including the full run of all three major rivers—the Yukon, Tanana, and Kuskokwim—plus a detailed rendering of specific tributaries on the upper Yukon, with links between the three watersheds. Like François Mercier, he appears to have combined a unique range of interests, abilities, and knowledge to play his role in the creation of the map that bears his name.[49]

While the border does not appear on the map, perhaps for reasons that differed between its Native and nonnative authors, it certainly must have had an influence on both Kandik and Mercier. The border has continued to influence both the oral and documentary records that survive to inform our understanding of them, their lives and times, and the subsequent history of their map. With its bold, simple outlines, linguistic diversity, and enigmatic symbols, the Kandik Map offers a dual "first-person" record of the geographical knowledge and interests of at least one Athabaskan man and one French Canadian active in the area during this time of enormous change, influenced and perhaps supplemented by Ivan Petroff and other northerners he met in 1880. It serves as a focal point for exploring the lives and times of its creators against the backdrop of the larger events unfolding in the region.

What can we learn about the map and its creators, and how? What does it mean? Information about the region, its people, and its resources was hard to come by in 1880 and as a result, sources about this period remain scarce to this day. Original documentary sources are limited in number and those that do exist circle around and about the same few people, places, and events so that a sense of the small community of neighbors and acquaintances grows rapidly from the firsthand accounts of missionaries, traders, and travelers. There are a few tantalizing drawings and even some rare photographs to provide a visual window on this world. Archaeological evidence provides clues to the expanding and enlarging trade activities, while ethnographies and oral histories document the travels and lifestyles of the Hän, Tanana, Gwich'in, and other Native peoples of the area. Oral traditions contribute some details about Athabaskan perceptions of the landscape, people, and places portrayed on the map, including stories recorded at the turn of the last century, and more recent accounts prepared for land claims research and heritage site and park developments.[50]

The provenance of the document now preserved at the Bancroft Library is also an important key to understanding the significance of the map and the human relationships that led to its creation. The Library consists of vast numbers of manuscripts, maps, photographs, and books related to northwestern North America originally collected by historian and publisher Hubert H. Bancroft more than a century ago, together with more recent acquisitions pertaining to the Pacific region and California. The story of how the Kandik Map came to be included in this library so far from its origins is uncertain,[51] although probably its acquisition by the library mirrors other patterns of trade and exchange of knowledge during those times. Its significance within the Bancroft Library and in the North are quite different, giving rise to reflections about the role of archives and libraries in preserving documents far from their original sources of inspiration and creation, together with the possibilities for preserving, losing, and regaining knowledge, and especially the cultural context related to documentary or other cultural heritage resources.

New electronic media technologies today offer many options for reproducing, manipulating, and sharing documents so that sources like the Kandik Map can be identified and reconnected to their original regions, where familial and community associations provide vital details for understanding them. Documents can take on a whole new life contributing knowledge and meanings to new generations. Copies of this map have been made in various formats and examined by First Nations Elders and younger people, linguists, researchers, and archivists, on both sides of the Alaska-Yukon border, at conferences and in more casual community settings.[52] The original map briefly returned to the North in 1987 when it was the subject of a heritage conference in Whitehorse on the pre–Gold Rush time of change. The map elicits different responses from people, depending upon their background, interests, knowledge of the landscape, and personal connections to the area.[53] These insights inspire the retelling of old stories and new opportunities for bringing people together to share their ideas across generations, cultural boundaries, and national borders. The information generated by these meetings has cast new light on the document, providing answers to some of its mysteries, illuminating its significance to people in different places, and pointing the way to new possibilities for further research.

Conclusion

The map is now over 125 years old, its documentation of the geographical information and other interests of its creators long since overtaken by later, more detailed maps produced by surveyors and cartographers, using increasingly sophisticated measuring and mapping devices. Yet its visual impact is still remarkable, especially when connected to the stories of the people who shared Kandik and Mercier's world. It has a special place in the mapping history of the North, both for what it shows and what it does not show as features and points of record. It is significant as one of a few extant northern aboriginal maps, drawn at a time of very active interest in the geographical knowledge of Alaskan and Yukon Native peoples. As such it represents an interesting echo of the current focus on gathering and preserving traditional indigenous knowledge and as a comparative document in the context of other indigenous maps worldwide.

The Kandik Map is a living archival record despite its age and the superceding knowledge of the world's spaces and places gained through satellite-generated digital imaging. The map continues to travel and to stimulate the exchange of stories across borders and between different cultural traditions and generations. It presents a special opportunity to examine the social life of a document over the course of more than a century, to consider changing concepts of landscape, land ownership, borders, and boundaries, and questions of cultural continuity, linguistic diversity, documentation and transmission of knowledge, and information sharing by and among northern peoples, while leaving many intriguing mysteries unresolved, perhaps forever. At the time it was drawn in 1880, just before the Klondike Gold Rush, the Kandik Map was the most complete picture of the major watersheds of interior Alaska-Yukon, representing the accumulated knowledge from Native and nonnative "story camps" in the region. Through its subsequent travels outside the North that knowledge

Elders Percy Henry from Dawson City and Ruth Ridley from Eagle Village share their knowledge of place-names during a Hän Literacy Workshop at the Yukon Native Language Centre in Whitehorse in December 2006. YNLC, André Bourcier, 2006.

was transmitted far beyond its origins. Through its preservation at the Bancroft Library it survived to return to the North, where it continues to generate new ideas about Kandik country, past, present, and future.

2

Searching for Paul Kandik

WHO WERE Paul Kandik and François Mercier? Did they work together on the Kandik Map, were other people involved, and how and where was it drawn? How did these men come by the knowledge they recorded, and why and for whom was the map drawn? The search for answers to these questions includes two very different types of sources—the oral traditions of Alaskan and Yukon Native people which originate in far earlier times, some of which have only recently been recorded, and the documentary sources compiled by nonnative people from the time of first direct contact between Natives and nonnatives in this area, starting in the 1840s. To date no documentary or oral source has been found to tell the story of the mapmaking process, leaving the map to stand on its own as the only evidence of that event. There is no handed-down memory of the map today either in the North among Athabaskan people who would be the descendents of Kandik or in Québec among Mercier's extended family. Knowledge of the map resurfaced in the Yukon and Alaska in the mid-1980s, when copies of it obtained from the Bancroft Library began to circulate, and in Québec as late as the summer of 2006, when it circulated among some of Mercier's descendents. Neither is there any clear record or memory at the Bancroft Library of the people and events surrounding the creation of the map or of the circumstances of its arrival in the collections there. The map itself does include a great deal of information and, when placed within the context of other oral and documentary sources related to its time, contributes significantly to understanding the larger picture of Native and nonnative collaboration and exchange at a pivotal time in the history of this region, while leaving its own story still shrouded in many unanswered questions.

The Kandik Map is the only documentary source found to date that specifically references Paul Kandik with these first and last names and the only one that links him with trader François Mercier. Searching for further information about Paul Kandik's identity involves tracing family and band groupings among the Hän Hwëch'in and other Athabaskan people living on either side of the Alaska-Yukon border along the Yukon, Kuskokwim, and Tanana rivers, including oral traditions, ethnographic and ethnohistorical sources, and archaeological and linguistic evidence. The name Paul survives among Hän Hwëch'in people and other groups as both a first and last personal name, though Kandik does not. Kandik does persist as a place-name meaning Willow River in English, and it appears to be attached to the same tributary of the Yukon on the Kandik Map and on current maps.[54] Mercier and Kandik may have been the first to record the name for the river, as the earlier Arrowsmith Map of 1854 attaches the name Antoine River to what appears to be this river.[55] The use of Kandik as a place-name probably predates both Paul the mapmaker and his map, but it may have been used as a personal name only for him and only within a narrow circle of nonnative people, perhaps for a very short time in the 1870s and 1880s.

Ethnographies and ethnohistorical sources provide further evidence of the relationships between Native groups, as well as the intricacies of Native and nonnative contact in this period, but none of them mentions Paul Kandik. François Mercier is identified in a number of sources recorded by his northern contemporaries, as well as many historical accounts since the map was drawn. He also wrote his own memoirs, which were translated from the original French version and published in the 1980s, a century after his departure from the North.[56] None of the period sources found to date written about or by Mercier mentions the map or Paul Kandik.

Neither Paul Kandik nor François Mercier, or their map is mentioned in any northern oral traditions found to date.

The map itself offers some clues pointing to various people and events contemporary to Kandik and Mercier that provide additional context for the map and its creators. Most important are the notes attached to and written on the map itself, which are not in Mercier's hand. Although these annotations are unsigned, an analysis of other manuscripts at the Bancroft Library confirms them to be the writing of Ivan Petroff.[57]

Map of Upper Yukon, Tananah and
Kuskokwim rivers
drawn by Paul Kandik, Yukon Indian.
lettering by François Mercier, trader.
1880.

Ivan

P.S. Best regards to the McIntyre
family. If you should meet a Mrs. Holman
at West Randolph tell her I left her
husband at Kadiak in good health, swinging
his broadaxe on the house he is building.
He is a very nice man. — I. P

Ivan Petroff worked as a U.S. government census agent in Alaska, traveling up the Yukon River as far as the Tanana in the summer of 1880. A comparison of his handwritten correspondence at the Bancroft Library reveals distinctive formations for "I," "M," "K," and other letters, identifying him as the writer of the label on the Kandik Map and perhaps its instigator or at least a witness to its creation. The Bancroft Library, Map of Upper Yukon, Tananah and Kuskokwim rivers 1880, Map Collection, G4370 1880 K3 Case XB; Petroff Collection, Letter of Ivan Petroff to his wife, May 10, 1881.

Petroff was a Russian émigré who spoke several languages fluently, including French, English, and Russian. His professional activities over three decades were a complex mix of highs and lows, with several episodes of dubious moral judgment, which ultimately ended his career. He enlisted and fought on the Union side in the American Civil War, and apparently deserted or disappeared several times during that conflict. He turned up in Alaska to work for the Russian-American Company in 1865, then moved to San Francisco in 1870, where he wrote newspaper articles for several years and was disgraced for writing falsified accounts of some events. His reputation either was not disclosed or not fully understood for he continued to find adventuresome work in the following years. With his knowledge of Russian and of Alaska, he was hired as a researcher by entrepreneur Hubert H. Bancroft. He traveled to Sitka in 1878 to copy and translate Russian-American Company archival documents, and wrote the chapters on the Russian period for Bancroft's *History of Alaska*, published in 1886. In 1880 he was hired by the U.S. government as a Special Agent to take the 10th Census in Alaska, and returned again in 1890 to oversee the 11th Census. Through his work for Bancroft and with the Census, Petroff was in close communication with many of the leading "outside" experts and interests connected to the North such as George Davidson of the U.S. Coast and Geodetic Survey in San Francisco and W. H. Dall in Washington. As well he met many of the indigenous people, traders, missionaries, prospectors, and other residents of the North during his travels and work.[58]

Petroff was considered a northern expert for many years, but his career came to an ignominious end in 1892 when it was discovered that he had altered the meaning of several Russian-American Company documents, which he was translating for the U.S. State Department. This was a very serious matter as the documents were tabled by the United States as part of its case in an arbitration of Bering Sea pelagic seal hunting rights, which were the subject of a major dispute between Great Britain and the United States.[59] Despite this tarnish on his reputation, Petroff's preliminary report resulting from his Alaska census work in 1880–1881,[60] the revised and enlarged version released in 1882,[61] plus Bancroft's *History of Alaska* are still recognized as some of the most important historical resources about the region, cited for many years as the first comprehensive American data on Alaska.[62] All three publications included maps that appear to have drawn on the Kandik Map for new information on the upper Yukon, and especially the upper Tanana and upper Kuskokwim rivers.[63]

Two stamped annotations on the original Kandik Map (one reading "Hubert H. Bancroft Map Collection, Bancroft Library" and the other "The Bancroft Library Map Collection") establish it as part of the early documents of the library, but the cataloguing records are silent on its origins and provenance. It is not possible at this time to determine whether Petroff gave the map to Hubert Bancroft to use in having the map drawn for his *History of Alaska*, or Bancroft or others acquired it for the library from some other source at a later date.[64] All of the available evidence points to Petroff as the likely instigator and possibly the only witness to the map's creation. Yet he mentions neither of the mapmakers in his published works or his correspondence, though he certainly met François Mercier and possibly Paul Kandik as well.

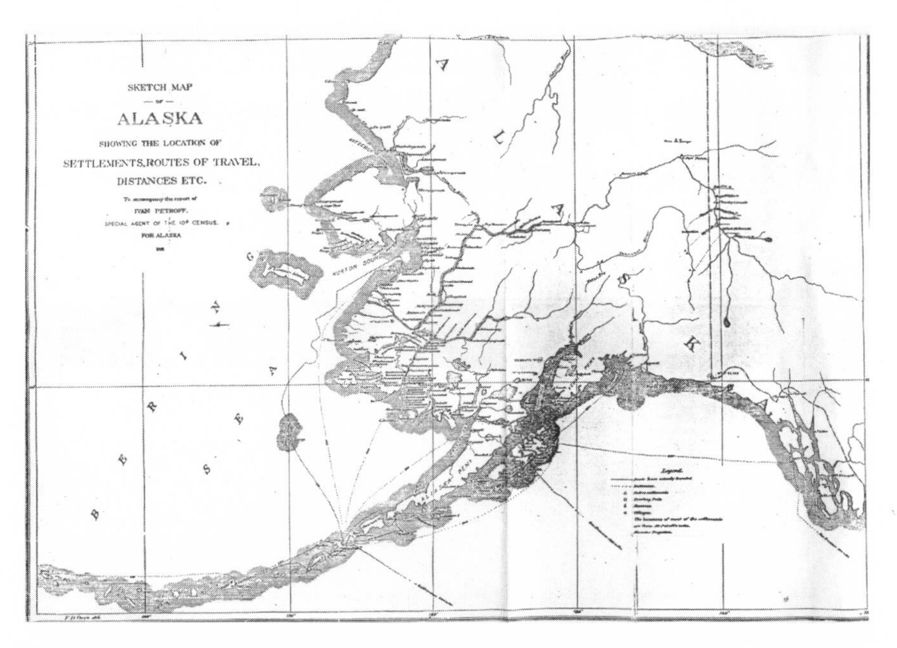

The Sketch Map of Alaska, which accompanied Ivan Petroff's first edition of his 1880 Census report published in 1881, appears to have copied tributary names and trail routes on the upper Yukon and Tanana rivers from the Kandik Map though he did not credit Kandik or Mercier. University of Alaska Fairbanks, Rasmuson Library, Alaska & Polar Regions Collections, Ivan Petroff, *Population and Resources of Alaska*, House Executive Document 40, 4th Congress, 3d session, 1881.

Taken together, all the oral traditions, contemporary documentary sources, and more recent research form an incomplete picture of the map and its makers. The gaps in the evidence about these people, and in tracing the provenance of the map, remain part of its mystery and pose more questions for further research. The scarcity of information concerning the two mapmakers and their map underscores the fluid nature of the times in which they lived, reflecting upon the contemporary local social structures, economic conditions, and relationships between Native and nonnative inhabitants along the Yukon River in the 1870s and 1880s. The disconnect between the map's creators and the documentary evidence left by their contemporaries, and between them and northern people today, is similarly a long-term and ongoing result of the massive changes and turbulent times, then and ever since, in this region.

Mapping an Identity

As the only source of information that is definitively related to Paul Kandik, what does the map reveal about him? Or more correctly, what did he portray of his identity, knowledge, and interests in his drawing that would lend detail to his life story? The map itself has two types of information—the "pictured" landscape and the "literate" names and other annotations. If the handwritten inscription attached to the bottom of the map, which declares that the "map" was "drawn" by Paul Kandik and "lettered" by François Mercier, is interpreted to mean that all the landscape features and other physical details were created and recorded by Kandik as the primary or sole "author," then he has documented a significant amount of information about his world and his travels. Mercier wrote the names for certain key landmarks in several languages, documenting other knowledge, contacts, experience, and interests that may or may not have been contributed solely by Kandik. Separating the physical "drawn" aspects of the map from the literate "named" features provides a useful method for speculating about the differences in perspectives and purposes between the two men, as Native and nonnative contributors to the map, as well as the possibility that other people had a hand in its creation.

The physical picture of Alaska-Yukon drawn by Kandik is grand in scale, scope, and detail. The map includes the three major tributaries of interior Yukon and Alaska—the Yukon, the Tanana, and the upper Kuskokwim—covering a landscape of tens of thousands of square miles and more than five thousand-plus miles along the three rivers. It represents an enormous range of geographical knowledge by any standard, especially considering the challenging terrain plus extreme climatic conditions year round, rudimentary modes of transportation in Kandik's traditional society, and the social conditions of his generation. How did Kandik gain the knowledge and ability to picture this vast area? Based on the accounts of other Native mapmakers such as Tlingit Chief Kohklux[65] and Nunamiut Simon Paneak,[66] who drew the landscapes through which they had actually traveled, it is likely that Kandik personally experienced at close range most, but perhaps not all, of the areas drawn. Given the 1880 date of the map, he could have depicted some areas on the basis of stories heard from other travelers, both Native and nonnative, and possibly from maps carried by some of them, which he may have seen and perhaps copied in his drawing.[67] His mode of river travel to these places almost certainly changed or at least expanded during the period of his lifetime, from the traditional birch-bark canoes expertly crafted and paddled by northern Native people in this area to include travel on the steam-powered riverboats introduced by American traders in 1869. These "fireboats" quickly transformed commerce in the region, enabling longer voyages in shorter times and opening up wage employment for Native men as pilots and laborers during the brief summer navigation season. Kandik's range and routes of overland travel may have altered during this period as well, since trade patterns and relationships among Native bands shifted along with the changing circumstances of the nonnative trading companies on the Yukon.[68] All of these factors could have influenced his ability to document such a large area with the trails and drainage relationships between so many river courses.

The map was drawn in pencil on a small sheet of paper (8.5 x 11 inches) and Kandik used these media to good effect in creating a variety of visual images of the region. The section of the map with the most intensive collection of landscape features and place-names is the stretch of the Yukon River between the White and the Porcupine rivers, all within the traditional territories of the Hän people along the Yukon River.[69] This is one of the most significant indicators, along with his name, of Paul Kandik's possible identity as a Hän man, or at least of his close association with people of this area. Kandik drew fifteen tributaries on the left and right banks of the Yukon River in this section of the map, starting at the White and proceeding down to the Porcupine River. Comparing his outline of these rivers, and the Native names attributed to them by Mercier, with more recent Hän Hwëch'in land claims, heritage, and linguistic research, it is possible to identify likely matches for most of these tributaries with traditional Native and current Yukon geographic place-names.

The tributaries drawn in this section are as follows (Mercier's names, when available, are shown first and indicated with an asterisk, with presumed current toponyms in parentheses, or presumed current toponyms alone when a river is not named by Mercier): the R. Blanche* (White River), the Ladue (unnamed by Mercier), Natchondé* (Stewart River), Chevreuil R.* (translates as Deer River in English, which was the name used by H.B.C. traders and others, now officially named the Klondike River), Tchandick R.* (Chandindu or Twelve Mile River), Thetawdé* (Mission Creek),

Ed Schieffelin's prospecting party and Native crew aboard the steamboat *New Racket* preparing to leave Saint Michael in 1882 for the upper Yukon River. The crew likely included Native pilots from communities on several sections of the river who could guide the boat through myriad channels and other navigational difficulties, as well as hunting and wood cutting for the prospectors on their first trip to the North. Alaska State Library, Wickersham Collection, Charles Farciot, SL-P277-017-002.

Tchandindou R. (may be Eagle Creek), Klévandé* (Seventymile River), Thégetondou ou Laroche R.* (may be the Tatonduc River), Tolkesettondou ou Bouillié R. (may be the Nation River), Kandick* (Kandik River), and the Charley River on which the name appears to have been erased. There

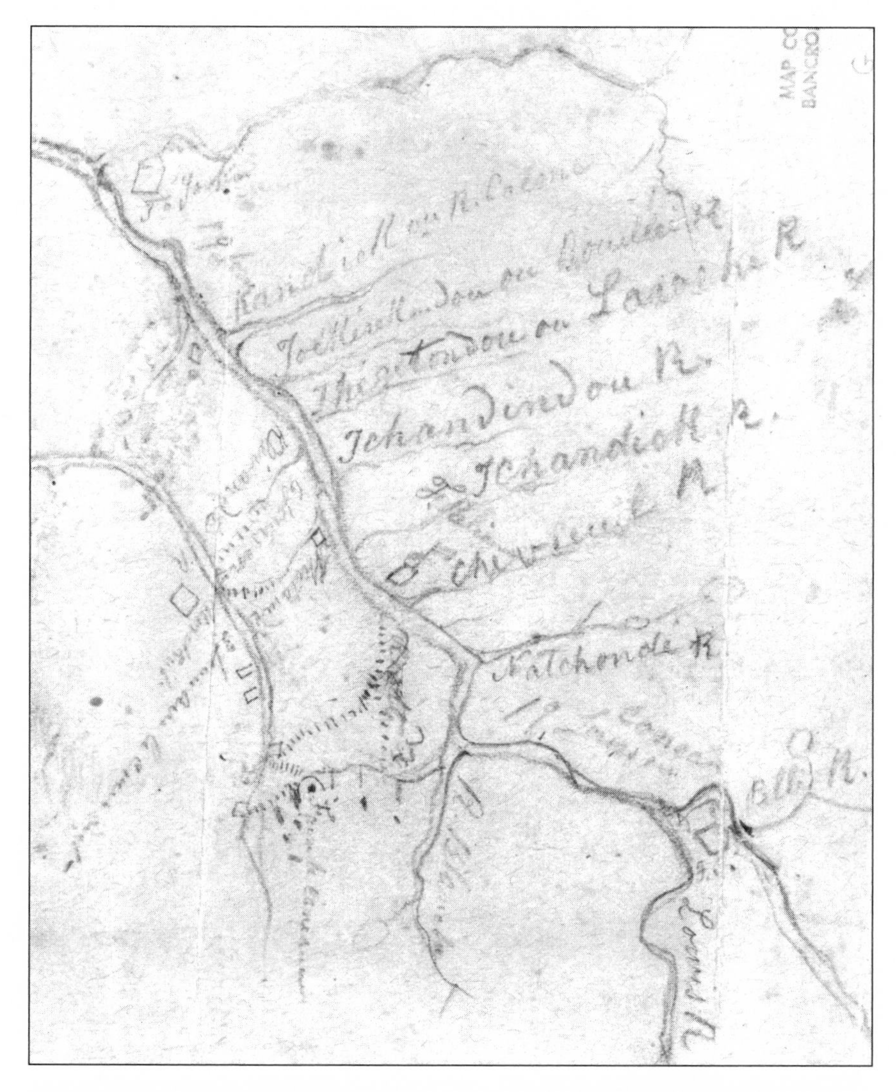

The most detailed section of the Kandik Map is the area in the traditional territory of the Hän Hwëch'in people on the upper Yukon River between the White and Kandik rivers, lending credence to the likely origins of Paul Kandik as a Hän man. The Bancroft Library, Map of Upper Yukon, Tananah and Kuskokwim rivers 1880, Map Collection, G4370 1880 K3 Case XB.

are two unnamed rivers upriver from the Klondike (which might be the Indian River on the right bank) and another one on the left bank. There is one surprising omission—Kandik has not shown the Fortymile River at all, perhaps reflecting its relative lack of importance in 1880 to him, his kin, and the local white traders. There is a V-shaped mark in the river at the confluence of what is now called Mission Creek (written by Mercier as Thetawdé*) which is likely Tthee t'äwadlenn (Eagle Bluff),[70] a prominent landmark that is the key identifying feature at the town of Eagle today and which figures in numerous Hän oral traditions. A quite distinctive, shaded and elbow-shaped bend with speckled marks along both riverbanks is shown north of Charley River in the area of Circle, Alaska, which reflects the many channels and islands in this stretch of the river.

The Porcupine River is shown entering the Yukon as a faint set of two lines with little detail and only one tributary on the south bank of the Porcupine extending down almost to meet the upper reaches of the east bank tributaries of the Yukon in this section, probably the Bell River. Curiously, no trails are shown between the Yukon and the Porcupine, not even the trail between the Tatonduc and the Bell, which had been heavily traveled in H.B.C. times.[71] There are also some large dark dots shown to the south of the Porcupine River, which might represent Native camps or perhaps some other landscape feature. The sparse detail shown for this northern region within Canadian territory may indicate the area either was not well known to Kandik or that it was not the focus of attention for the intended audience for the map, which may have included the American trader/prospectors on the upper Yukon, as well as Ivan Petroff.

From the confluence of the Porcupine the Yukon is drawn as a gradual set of lines flowing downstream in a westward direction ending at the left side of the page. A round, fairly large circle shape attached to a short tributary on the left bank downriver from Fort Yukon is another prominent feature, likely a fishing lake of some note to be so distinctly drawn. This may be the fishing lake noted in early traders' accounts[72] and possibly is Birch Lake as shown on maps today. Further downriver there is a prominent bend shown on the right bank a short distance above the Tanana, likely the bottom of the Rampart and often noted as one of the camps of the powerful Chief Shahnyaati'.[73]

The Tanana-Yukon confluence is drawn as a fairly large V-shaped delta. From there the Yukon continues as a significantly broader double outline to the west, with less detail and fewer tributaries, mostly unnamed. The riverbanks are distinctively emphasized with shaping and shading, indicating that Kandik probably had some detailed knowledge and memory of the river itself. There is no detail shown for physical landscape features beyond the river, such as lakes or mountains, from the Tanana down to the lower end of the Yukon River at the left edge of the paper. Likewise the drawing of the southern portions of the Upper Yukon (Lewes R.*) and Pelly (Pelli R.* as named by Mercier and mistakenly placed on the MacMillan River tributary of the Pelly) rivers, and the westernmost portion of the Kuskokwim River, is faint in contrast to the more central Yukon areas and contains little detail.

The Tanana River resembles the drawing of the Yukon in the strength of its two main lines and shadings, flowing in a soft curve between the Yukon and the Kuskokwim. The upper Tanana is shown as a single line curving slightly to the east and then south. There are some prominent bends shown in the mid and lower ends of the river to the north. Only four tributaries are shown and only three are named—the largest may be the Delta River or Tok River today (named Tutluk* on the map, but possibly written in Petroff's hand) entering in the middle section on the left bank; the next one downriver on the right bank is probably the Salcha River (Saklikegata,* also possibly written in Petroff's hand). Another which is unnamed on the left bank is probably the Kantishna River, with one tributary shown on its south bank likely being the Too Tlat or Toclat River.[74] There is also a short unnamed tributary with a round lake attached to it on the right bank close to the Tanana-Yukon confluence, another fishing lake, probably Minto Lakes as named on other early and current maps.[75] It appears that the lettering of the names on the Tanana tributaries is Petroff's and not Mercier's hand, suggesting that this information came from others besides Mercier and Kandik, perhaps traders Arthur Harper or Jack McQuesten, who had traveled on the lower Tanana by 1880. Overall the Tanana is rendered as a much simpler drawing than the section of the Yukon in Hän territory, possibly reflecting less knowledge on Kandik's part, and perhaps gleaned from fewer and faster trips in this area. This may be another indicator that Kandik was not a Tanana River person by birth or residence, but had traveled and traded in the area, either before or after white traders arrived, and perhaps both. He also could have had knowledge of the area from oral traditions passed on by Tanana people when they came over to the Yukon to trade, or shared between intermarried Hän and Tanana families.

The Kuskokwim River is shown as a strong set of two single lines on its upper reaches that form a Y shape near the Tanana on the northeast, joining together at a fork which then becomes a wider double-lined course stretching to the southwest on the lower left edge of the map. One unnamed tributary below the forks is shown flowing from the south. Both this tributary and the main river below are drawn as lighter sets of undulating lines, again possibly indicating that Kandik was less familiar with this part of the Kuskokwim, or indeed was drawing an area that was completely outside of his personally traveled landscape and riverscape knowledge.

Although river drainages and lakes appear to be the main focus of Kandik's drawing there are some other notable physical features. Extensive shadings that appear to represent mountains are drawn in the middle section of the map, surrounding the tributary of the Tanana named as the Tutluk,* and the upper reaches of the two forks of the Kuskokwim. These likely represent the Alaska Range mountains that dominate this region.

Charles Farciot photographed some Tanana men who had traveled from the Nabesna area to the Yukon River near François Mercier's Belle Isle post c. 1883. Alaska State Library, Wickersham Collection, Charles Farciot, SL-P277-017–020.

To the southwest of the upper end of what may be the Kantishna there is a group of "mountain" shadings and on the edge a distinctive set of three dark lines, perhaps representing the giant peaks of Denali (Mount McKinley). Kandik could have heard stories about that huge landmark during trips down the lower Yukon, and if he traveled on the Tanana, he could have seen Denali, the beacon of northern travelers in this region for countless generations.[76]

At the bottom right corner of the map an elongated and shaded shape at the end of a river marked as Chilkat R.* by Mercier appears to be a representation of the Lynn Canal and the tidewaters of the Pacific Ocean. This feature is lightly drawn and again the strength of the representation may indicate a more tentative knowledge of this route and region—perhaps known to Kandik from oral traditions shared between his people and more distant Native trading partners. Tlingit traders from Klukwan and other coastal villages paddled downstream from the upper Yukon, probably as far as White River, and perhaps further in earlier years, but interior Athabaskan people were not traveling to the southeast coast until a few years after this map was drawn. Some details of places and routes Kandik depicted on these "outer edges" of the map were probably gleaned from nonnative sources too, since various traders and travelers had been in the region for over thirty years. Robert Campbell traveled by canoe from the Pelly River to Fort Yukon in 1851, and Ketchum and Laberge retraced his journey upstream and back in 1867.[77]

As well as the physical landscape features, Kandik included a number of other types of markers that seem to indicate human occupation and activities or travel. On the Yukon there are seven prominent markers, which appear as square boxes with flagpoles and flags atop them, which represent some of the most prominent trading posts of the fur trade era to 1880. These posts were labeled by Mercier as F. S.* (undoubtedly Fort Selkirk at the Pelly-Yukon confluence), F. Raliance* (Reliance in English), F. Yukon* (at the Porcupine), an unnamed post just upstream from the Tozitna (probably Fort Mercier) and F. Tanana* just below that river, Nulato F.* shown downriver from a large unnamed tributary that is certainly the Koyukon River, and an unnamed post just above the Anvick R* (Anvik River). These posts were all established by nonnative traders starting in Russian-American Company and Hudson's Bay Company times in the early to mid-1800s, and continuing with the activities of American traders after

1869, including the Alaska Commercial Company and others for which Mercier worked. Kandik may have drawn these forts or perhaps Mercier did, as they document some of the posts he established and places where

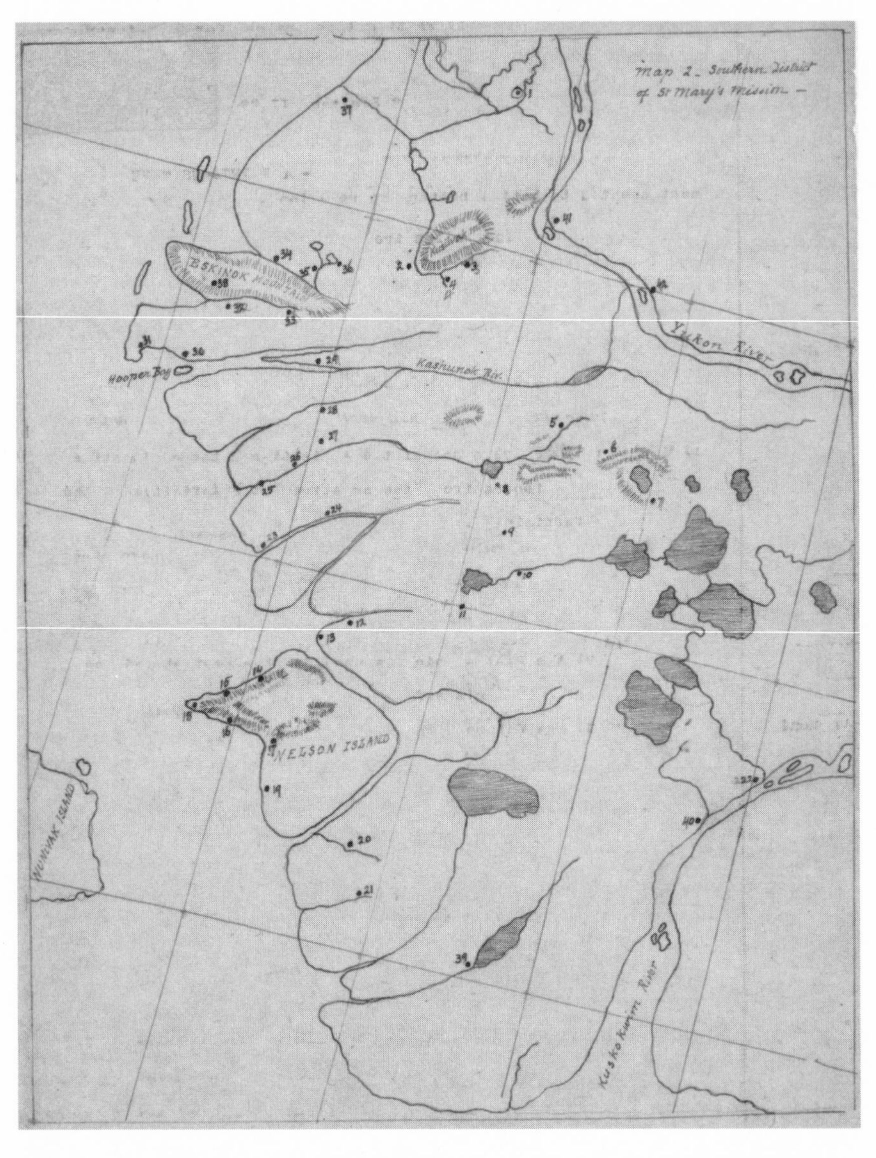

These two maps were produced from information and drawings provided by Native Alaskans to Father Joseph Jules Jetté, a Jesuit priest who recorded Native Alaskan languages and place-names along the Yukon River in the early 1900s. Jesuit Oregon Province Archives. 503.21a Jetté "Map 1" of the Yukon River delta area, n.d.; 503.22a Jetté "Map 2" of the Yukon River delta area, n.d.

he lived from 1868 to 1885.[78] In the top left-hand corner there is a faint drawing of what appears to be another larger fort and maybe some large boats, plus some marks that could denote the route from the lower Yukon to Saint Michael's Post located on the Bering Sea coast.

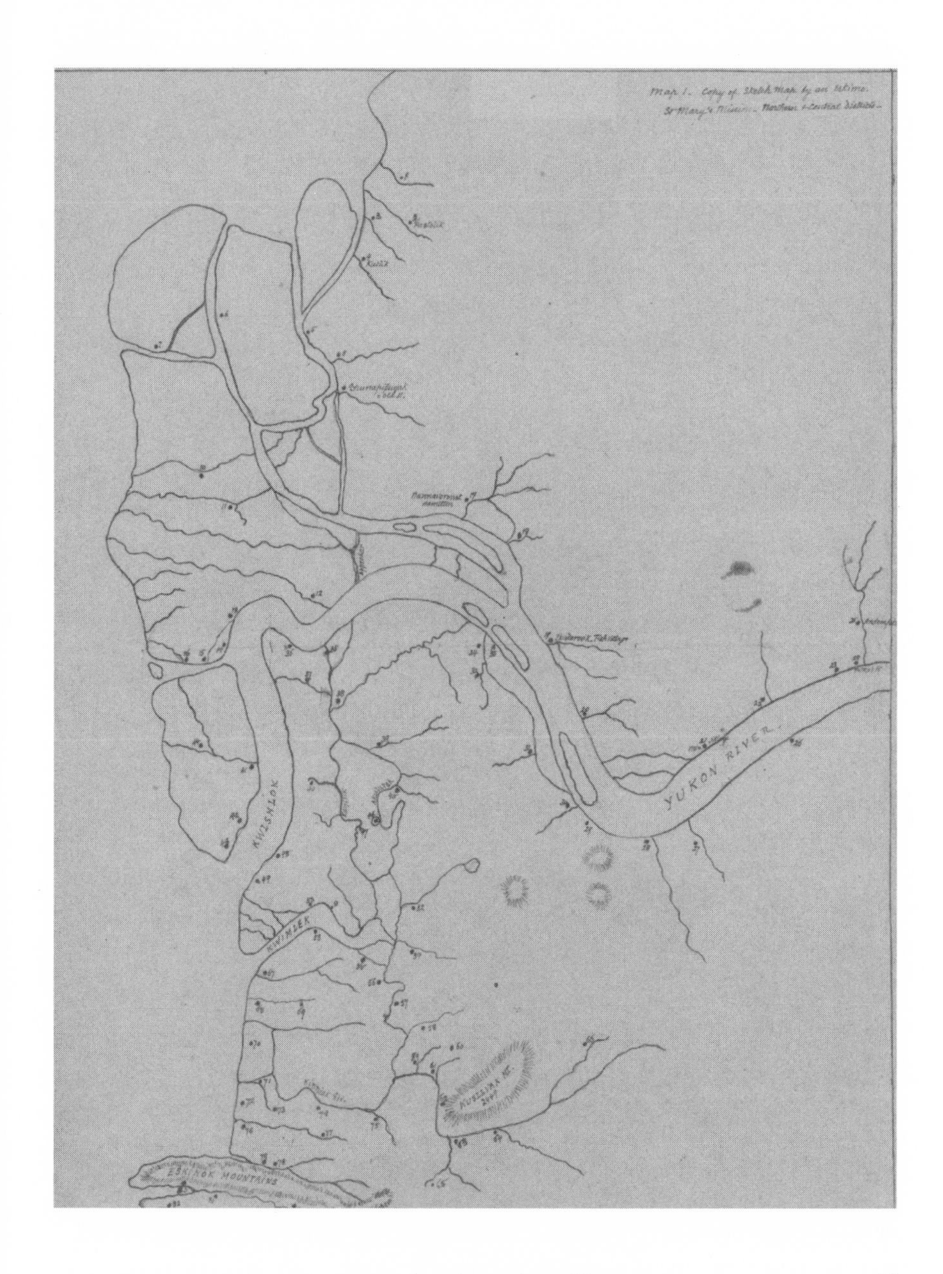

The depiction of the Yukon River ends on the left edge of the paper just downriver from a trail between the Yukon and lower Kuskokwim, but does not show the Yukon delta or any part of the Pacific coastline. This could mean that Kandik never traveled all the way to the coast or that Petroff used other sources for this part of the Yukon River and suggested the western limit to the drawing. As Dee Longenbaugh has noted, this area was extensively traveled, mapped, and named by Russian-American Company employees and Creole associates in earlier years.[79] Petroff would have seen their documentation at Sitka and also may have had maps drawn by Native people in this area similar to the Kandik Map.

There are some smaller square shapes and dots, which may denote Native dwellings or meeting places familiar to and drawn by Kandik. On the Yukon there is a small square without a flag drawn on the left bank at Thetawdé* (Mission Creek), the site of the Hän camp known as David's Village (now Eagle Village), and also of one of the posts established here by Mercier in the summer of 1880 called Belle Isle.[80] Downriver and across from the Kandik River on the left bank there are two small squares, located at sites reported to be Hän camps occupied by Chief Charley's Band during the mid- to late 1800s.[81] On the upper Tanana there are four small square shapes that probably represent camps of some type, likely the dwellings of Upper Tanana people since this map predates white settlement there.[82] Another small square located on the west bank on the lower part of the river may be a Native meeting place or camp, or perhaps the cabin of Mr. and Mrs. Bean, the first white family in the area.[83] It is located on the opposite bank but in approximately the same place documented on his 1885 map by Lieutenant Allen as Harper's Station,[84] and named "Short Station" by Jack McQuesten, referring to the same place, which is where he left Arthur Harper to establish a new trading post on their first steamboat trip up the Tanana in 1880.[85] On the lower branch of the upper Kuskokwim River there are two more square shapes that may indicate Native dwellings or camps, or early white trading posts on that river.

The map has a number of other features that appear to document travel routes, trails, and perhaps portages or other transportation-related information. There are six series of hatch marks that likely denote canoe routes, portages, and foot trails. One set connects between the point on the left bank of the Yukon across from Fort Reliance, which would be Catsah's camp or Nu-kla-to[86] as shown with various spellings on later maps, and the Tanana. There is a faint word written across the marks at this point, which is obscured by other marks but might be “Katcé . . .*”[87] and is probably Mercier’s spelling for the chief referenced elsewhere as Catsah who had a camp there. He was well known to Mercier and other traders, and was one of the Hän men who piloted Mercier upriver to establish Fort Reliance in 1874.[88] The trail from this point splits into two forks, one leading to the Sixtymile, the other to the upper Tanana. A second set of marks connects between the upper section of Thetawdé* (Mission Creek) and a point on the upper Tanana located a short distance down from that trail. A third set of marks connects between Klévandé* (the Seventymile River) and another point a little lower down from the second set on the Tanana. Two more sets of hatch marks lead from the upper end of the Sixtymile to different points on the uppermost reaches shown for the Tanana. These marks may indicate the number of days for traveling by canoe and on foot between these river valleys. They closely resemble the Native trails and portages noted in the oral traditions of Hän and Upper Tanana people, and reported by various early nonnative travelers, as well as later ethnographers.[89]

There are two other series of shapes that look like exclamation marks that may denote foot travel; days of journey by canoe, portages, and stopping points; or some type of trail markers. One set appears on the upper Sixtymile River connecting to the southern end of the upper Tanana. The other set of exclamation marks forms a V-shaped line extending from the Tanana tributary marked Tutluk* to the headwaters of the southern branch of the upper Kuskokwim and southwestward along that tributary almost to the fork of the river, then cutting across land to the midpoint of the northern branch of the Kuskokwim, crossing it and continuing across land to the upper end of the tributary shown at the lower end of the Tanana River, and ending almost at the confluence of the Yukon and the Tanana. The Kuskokwim trails had not been traversed by any nonnative people by 1880,[90] so this information must have come from Kandik’s personal knowledge, or other Native travelers and oral traditions.

There is a different style of faint hatch marks and lines with four dots connecting between a shape resembling a long lake, which might be Lake Laberge, on the southern end of the Louis* or Lewes R.* (the upper Yukon), and the faintly drawn river named as the Chilkat R.* at the very bottom lower-right section of the map. This represents the trading route of the Tlingit people from Klukwan and Dyea on the coast who traveled to the interior several times annually and traded with interior Tutchone people, who in turn traded with the Hän people. This trail was well known to Petroff and others by 1880 through the maps of George Davidson, who incorporated information shared by Tlingit Chief Kohklux.[91] A similar style of dots is depicted for the portage between the lower Yukon and Kuskokwim below Anvik. The three distinctive styles of trail markings may indicate that several different people drew these features on the map. Kandik may have drawn the trails between the Yukon and the Tanana in Hän country, but it is unlikely that he personally traveled on all of the other trails in such widely dispersed areas prior to 1880.

Two other symbols appear beside an unnamed river, which is probably the Sixtymile River. One is a dot with a half circle over it and some indecipherable words beside it, located in the middle of the exclamation marks on the Sixtymile trail. The other is a shape like an inverted "M" located on the east side of the hatch marks between Katcé* and the Sixtymile. Ferdinand Schmitter recorded a few trail markers commonly used by Hän people at Eagle in 1906 and these symbols may possibly be a similar type of marker.[92] Further research with Hän people may reveal some additional information about these symbols.

It is possible that some of these indications of human activity, which document "trails," "routes," and "dwellings," were not drawn by Kandik on the map. Its 1880 date of creation postdates the earliest reported travels of nonnative traders in the upper Yukon, the lower Tanana, and the Sixtymile valleys, so it could be that Mercier or other nonnative traders such as Harper, Mayo, or McQuesten, and their Native wives and relatives, supplied information about or drew some of these details, perhaps including what was learned from Russian-American Company and Hudson's Bay Company traders before them as well.[93] Kandik could also have seen copies of maps of the region carried by nonnative travelers and traders, or sketches drawn by them, as Harper and others were reported to have had published maps in their possession at this time. Petroff could have brought a copy of Dall's book or a sketch of his 1875 map with him. He reported that he had consulted "all available" reports and maps before leaving on his trip.[94] So Kandik and Mercier's map could have been a collaborative effort including a number of Native and nonnative people, and it certainly included information learned from others, given the broad range of territory depicted.

Oral traditions and ethnographic and archaeological research related to the Hän people all provide evidence of extensive and intensive travel and occupancy along the Yukon between the Sixtymile and the Porcupine rivers, plus use of the trails and routes noted on the map to travel to the uplands between the Yukon and Tanana to hunt, and by both Hän and Tanana people to go between these places for trade. This concentration of detail in Hän areas supports the idea that Paul Kandik was a Hän man.[95] However other areas represented on the map cover much more than this traditional Hän territory, from the Chilkat River to Anvik on the Yukon, the whole length of the Tanana, plus the upper reaches of the Kuskokwim, indeed virtually the whole interior of the Yukon and Alaska. There are numerous reports by nonnative observers of both individuals and groups of Native people traveling long distances beyond their usual locales to trade and visit. In particular the H.B.C. traders at Fort Yukon and Fort Selkirk,

Reverend Vincent Sim worked for the Anglican Church Missionary Society and traveled on the upper Yukon River in the summer of 1883 teaching hymns plus rudimentary reading and writing lessons to Native people. Here he is preaching to Hän people at Fort Reliance. Alaska State Library, Wickersham Collection, Charles Farciot, SL-P277-017–021.

and Russian-American Company traders on the lower Yukon and Pacific coast, provided powerful reasons for aboriginal traders to alter their traditional trade and travel patterns, offering attractive new metal implements, colorful calico cloth and beads, guns, ammunition, and other useful items. The new trading posts were also places for social gatherings and intellectual exchange—a place to observe new customs, learn about new ideas such as Christian teachings, and gain essential skills like reading, writing, and arithmetic required to compete in the white man's world.[96]

While it is clear that Native people traveled extensively in this region, both before and after the arrival of white traders, the full scope of the Kandik Map is well beyond the areas an individual Hän man like Paul Kandik could have traveled prior to white contact, or in the Russian American and Hudson's Bay Company period. The range of aboriginal precontact and early historical period travel was affected by a number of factors including seasonal weather constraints, the substantial time commitments of subsistence hunting and gathering, the physical limitations of canoe and foot travel, and the territorial control maintained by different chiefs or leaders over particular areas. Although there were no borders or surveyed boundaries, these territorial distinctions were sometimes fiercely contended. Oral traditions recounted by Johnny and Sarah Frank,[97] Simon Paneak,[98] and Adeline Peter Raboff,[99] as well as many nonnative accounts, document stories of battles and killings between different Native groups in the Alaska-Yukon interior, the result of competition for resources and maneuvering to gain economic power. Native guides often related these stories to nonnative travelers as the reason for refusing to travel beyond their own familiar territory, as recounted by Ogilvie[100] and many others.

The broad scope of the physical terrain shown on the map could mean that Kandik was traveling with nonnative people beyond his traditional territories, protected from hostilities with neighboring Native groups by the mystique and economic utility of the new players in the changing trade networks of the region. He could, as well, have been traveling farther and faster with the advantages of the new steamboat technology, and for the purpose of engaging in new employment opportunities as guide, hunter, river pilot, and laborer, introduced by American traders after 1867. He also would have been immersed in the oral knowledge of landscape shared among kin and between neighboring groups of aboriginal traders for countless generations. While not definitive as evidence, the "pictured" landscape and riverscapes, identified by Petroff as being drawn by Paul Kandik, especially the detailed rendering of Hän home territory, trails, and place-names, plus the broader geographic representation beyond traditional Hän territory, together with the date of the map, all lend support to these conclusions about his probable identity as a Hän man, well acquainted with the oral traditions of his aboriginal heritage, expanded through the new economic opportunities presented by nonnative traders, and possibly the maps carried by travelers from outside during this time of change.

Words and Sounds: Reading and Listening for the Past[101]

The "picture" of the physical landscape that Kandik drew is joined with Mercier's "literate" annotations on the map in the form of place-names, to create a multilayered representation of the world they knew. The place-names "lettered" by Mercier, when spoken aloud, introduce sound to the visual information, adding a significant dimension to the impact of the map and more mysteries to ponder. The Native place-names he wrote may or may not have been contributed by Kandik, so they could be further evidence of Kandik's identity, or perhaps not. Mercier also documented his own linguistic diversity, and probably some knowledge handed down by previous traders from H.B.C. times, with both French and English place-names recorded on the map.

More information is provided by Petroff's handwritten annotation located on one corner of the map reading "Rec'd from M. Mercier at St. Michael's, July 1, 1880." The other Petroff annotation designates Paul Kandik as a "Yukon Indian," a verbal clue to his identity that suggests a number of possible origins for him, while the combination of Paul and Kandik offers more ideas for consideration. François Mercier's name is also significant as evidence of the dynamic cultural mix in the times and places represented by the map, although he is designated by Petroff simply as "trader." In this same note, Petroff entitled the document "Map of the Upper Yukon, Tananah and Kuskokwim rivers," providing an indication of his viewpoint concerning its focus and significance. It appears as well that Petroff may have written some of the names on the Tanana section of the map himself. The place-names, personal names, date, and other annotations relate to a small number of closely linked oral traditions and documentary sources that further illuminate the identities of the map-makers and the significance of their map.

What does Petroff's inscription that was attached to the bottom of the map identifying Paul Kandik as a "Yukon Indian" mean? To carry out his census work in the summer of 1880 Petroff traveled up the Yukon with trader Leroy McQuesten on the A.C.C. steamboat *Yukon* as far as Nulato,[102] then continued upriver by canoe with Native guides all the way to the Tanana. Along the way he met and counted Native and nonnative people. In the first preliminary edition of his report, *Population and Resources of Alaska*, transmitted to the House of Representatives Committee on the Census in January 1881, he refers to all the Indians of the interior of Alaska as "the Ingaliks, or the People of the Great Interior," sometimes listing group names and at other times only places where he met people: Nuklukayet 2 Whites, 27 Indians; Village above Ramparts 0 Whites, 110 Indians; Fort Yukon 2 Whites, 107 Indians; Gens de Large 0 Whites, 120 Indians; Tennanah River 1 White, 700 Indians; Koltchones, roving between Yukon or Kuskokwim 75 Indians. He added that "on the Yukon River, above the fort of the same name, we know of the following people trading with Americans at Fort Reliance, who may be on British soil: Charley's people 48, Fetoulin or David's people 106, Fort Reliance (one white) 82."[103] He commented that most of these people had become familiar with white traders over the past fifteen or twenty years, adopting "parts of their dress." The exceptions were the more remote people of the unexplored Tanana who came down to the Yukon to meet the traders and numbered between 650 and 700. He noted that "last summer a trader's steamer pushed up this river with the intention of going some 200 miles; how far they did go we do not know at present."[104]

In the longer edition of his report published in 1882 Petroff revised his description of the interior people, calling them "the Athabaskans" and citing W. H. Dall as the authority for the group names he used.[105] These were the Natsit-Kutchin, "known to the traders as gens du large" who lived on the Porcupine, the Han-kutchin "known to the traders as gens des faux" (*sic*, probably a misprint as the usual term was the French word *fou*, apparently used by early voyageurs because of the energetic dancing of these people) who "trade with the natives of the Copper River and those of the upper Tanana River," and the "Yukonikhotana" or "men of the Yukon" who lived and traded from Fort Yukon west to Nulato but preferred to "assemble at Noyakaket and Nuklukaiet."[106] In adopting Dall's terms for this longer version of his report, Petroff followed the names commonly used by nonnatives for several decades to recognize distinctive groups with their different languages, styles of clothing, modes of travel, and traveling areas. These names and descriptions varied little over the years from the first H.B.C. traders' reports. Whymper, who had traveled with Dall upriver in 1866, listed "the 'Kotcha Kutchin' or lowland people, the Indians of the immediate neighborhood . . . divided under two chiefs . . . dwell on the Youkon itself and on Black River. . . . 'An-Kutchin' tribe who dwell higher on the Youkon (or Pelly, as it has sometimes been called at this part of the river) and are known by the voyageurs of the Company by the flattering epithet of 'gens de foux.' The Tatanchok Kutchin tribe or 'gens de bois', from the upper Youkon, the 'gens de bouleau' or Birch river Indians, and the 'gens de Rats' or natives of the Rat or Porcupine River [and] . . . the Tananas, or 'gens de butte', the 'Knoll people', or mountain men . . . [who]

Ivan Petroff was an official U.S. Census agent in Alaska in 1880 and again in 1890. In this undated photograph he was writing down information at a Native village, possibly on the lower Yukon River. The Bancroft Library, Petroff Collection, C-B 989, folder 6.

were the most primitive."[107] Raymond used similar names to describe the people he met in 1869 at Fort Yukon. Only a few writers used the term "Youcon" Indians, notably H.B.C. trader Alexander Hunter Murray,[108] as well as Anglican missionaries Robert McDonald[109] and Vincent Sim,[110] primarily referring to people at Fort Yukon or farther up the Yukon. So Petroff's inscription "Yukon Indian" on the map may mean that he understood that Paul Kandik either originated from upriver of Fort Yukon, supporting the idea that he was a Hän man, or was a "Yukonikhotana" from below the fort, linking him more closely with people who traded at the mouth of the Tanana.

At the very least "Yukon Indian" places him within the context of just a few particular northern Athabaskan communities of this era, from which details of his life can be inferred. All of these groups lived a seminomadic existence, moving through a seasonal round of locations to harvest food and furs, adding new trade rendezvous to older meeting places when European fur traders arrived in the region. He would have traveled from earliest childhood days by birch-bark canoe along the rivers and on foot over the land routes depicted in the central areas of the map. In precontact times, and in the early fur trade era, Kandik's regular travels would have been limited to areas linked to his particular band's seasonal rounds and trading activities, which could have extended over several hundred square miles, though the earliest traders reported that some individuals or small groups made much longer trips to trade.[111] After the introduction of steamboats to the Yukon River in 1869, Mercier plus other traders utilized this new technology to push their trading activities further up the Yukon River in the mid-1870s[112] and to explore the lower Tanana in 1880[113] and later, with the help of Native pilots and crews.[114] Paul Kandik may have been one of the Native men who made the transition to travel and employment on these boats during the 1870s, which could account for the broad geographical scope of his drawing in 1880.

Another method for trying to identify Paul Kandik's origins is to analyze the place-names on the map, the largest number being in the Hän traditional areas. The number of languages recorded by Mercier on the map is indicative of his facility with several local languages. For most of the tributaries he attached names derived from Native place-names, but he also provided French and English names for some of them. The names illuminate the challenges facing early traders and their Native neighbors in communicating with each other given the complex multicultural and multilinguistic demographics of the area during these years. The names as written by Mercier represent his "hearing" of the complex sounds of the Athabaskan languages of the area, spoken by people from more than one linguistic group, colored by his French Canadian mother tongue and English as his second language. Some of the names were probably passed on to him or his fellow American traders by H.B.C. men before they moved up the Porcupine, and by some of their interpreters, or former Native customers who stayed behind on the Yukon and started trading with the new American companies. Some may reflect one of several Hän dialects for this area, while others may reflect the Gwich'in language of Fort Yukon or Porcupine River people. Still others on the Tanana may represent the names used by people in those areas or at least Kandik, Mercier, or Petroff's rendition of those names.

In recent decades linguists have recorded place-names with many groups of Athabaskan speakers from Yukon and Alaska, which provides a basis for analyzing the Native place-names on the map. Although there are many similar words and phrases in the closely related Native languages in this area, there are also distinctive differences in the way key word endings are constructed between different groups. For example, one ending commonly heard in association with river or creek names in Hän, Gwich'in, and Upper Tanana means "along the course of something," or "along a river or creek course." In each language the ending appears in a different form. In Upper Tanana it is heard variously as "niik," "niig," or "niign," as in "Naabiaa niign" or "Naabia River" near Northway, Alaska, at the Alaska-Yukon border. In Gwich'in this ending is heard as "njik," as in "Teetl'it Gwinjik," the headwaters of the Peel River, or "Nagwichoonjik," the Mackenzie River. Typically when the ending "jik" is attached to a name, it is a Gwich'in speaker's version of a river or creek name. Among Hän speakers there are some distinctive variations in the way this ending is spoken between people originating at Eagle in Alaska and others from the Klondike in Yukon. People from Moosehide near Dawson City refer to the Klondike River as "Tr'ondëk" or "Tr'odëk," while Louise Paul, born c. 1921 in Dawson but raised in Eagle, spoke of "Tr'oju" and Chief Isaac, who originally came from Eagle and married a Klondike woman, was quoted by journalist Tappan Adney around 1900 as naming his people "Tro-chu-tin."[115] So the two Hän endings vary from each other as well as differing from the Gwich'in names. Another example of these variations is the name for the Fortymile River, recorded by Mary McLeod from Moosehide as "Ch'èdahndëk," while a Gwich'in speaker's version would be "Ch'dehnjik." How do these variations relate to the names on the Kandik Map and what might they signify as to Paul Kandik's origins? Or do they indicate that a variety of different speakers served as Mercier's sources for these place-names?[116]

Of most significance to Paul's identity, Kandik is spelled with the ending in a form most closely resembling a Klondike River Hän ending, as "dik." The Gwich'in version would be "K'àii njik," "K'àii" being the word for willow, and "njik" for a watercourse, translating to English as Willow Creek. The late Willie Juneby, another speaker from Eagle, named this same tributary in the Eagle Hän language as "K'ày' ju."[117] Assuming that Paul was associated with a pronunciation of Willow Creek that resembled his own spoken tradition, he may have been a Hän man originating from the Klondike and later living at the Kandik River. In this connection it would be important to be able to determine when the name Kandik was first recorded as the designation for the river. Currently the Kandik Map is the earliest source found for this place-name.

Looking at the other tributaries named on the map, Mercier writes three of them with very distinctive "dou" endings for the three rivers on the east bank of the Yukon south of the Kandik. These names may represent his attempts to capture Eagle speakers' "ju" endings for the names. The Tolkesekandou* may be the Nation River as named on current maps, though it does not resemble any Hän names recorded recently, while the Thegetondou* may be Mercier's rendering of some name for the Tatonduk on current maps. The Tchandindu* may be his hearing of some Hän name for Eagle Creek, but it does not resemble the name recorded by Willie Juneby with linguist John Ritter in the 1970s.[118] Although it appears to resemble today's toponym Chandindou, or Twelve Mile Creek, it is more likely that the name Tchandick* appearing as the next tributary on the east bank upriver on the map is the Chandindou. If this is correct then it appears Mercier has heard a Klondike River speaker's version, which would usually have a "dek" or "dik" sound as the ending for the name, rather than the Eagle version with "ju" and rendered as "dou" in today's toponym. It is possible that the current toponym Thane Creek is derived from this name.

There are three names written by Mercier with a "dé" ending. Klévandé* closely resembles the Eagle Hän name for Calico Bluff, which is Tl'evär tthèe', located on Tl'evär juu, now called the Seventymile River in English. It is possible that Mercier attached the Hän name for the bluff to the river, or alternately he may have been adapting the name to resemble a name in his own French Canadian linguistic tradition. The name attached to Mission Creek on the map is Thetawdé,* which resembles an Eagle speaker's name for it, Tthee t'äwdlenn juu, which incorporates the name for Eagle Bluff at the mouth of the creek, known as Tthee t'äwdlenn. Finally Natchondé,* which appears above the Klondike and is certainly the Stewart River of today, resembles the Northern Tutchone name used by Mayo people for that river.[119] Mercier has included an acute accent on the final "e" of all three of these names ending in "de." His "dé" ending on river names presents another puzzle, perhaps indicating Mercier's hearing the words as a similar pronunciation to a French speaker's version, or that he adapted the Native names to his own mother tongue.

On the Tanana only three names are included, besides the Tanana itself, which is the only one that appears to resemble the rest of the map lettering purported to be Mercier's work. The Tanana tributary names appear to be written in Petroff's hand as noted previously. On the left bank just above the confluence at the Yukon is Nuylakaut,* which may be his version of the name for the Native meeting ground and trading site at the Yukon-Tanana confluence, written by other observers as Nuklukayet and many

Charles Farciot photographed this group of women and children at a cemetery, most likely at Nuklukayet c. 1883. These may be the families of the traders Harper, Mayo, and McQuesten, who were living there at the time. Alaska State Library, Wickersham Collection, Charles Farciot, SL-P277-017—030.

other forms.[120] People from various language groups would have used a similar name for this important rendezvous site, each with their own distinctive linguistic form. One recorded version in the *Koyukon Dictionary* is Nuyhakaut, meaning "something coming together," which is descriptive of a confluence.[121] Petroff traveled upriver with his Native paddlers as far as the Tanana trading post in 1880, where McQuesten, Mayo, Harper, and their Athabaskan wives were then living. So Satejdenalno (Katherine "Kate" James McQuesten), who could speak Russian as well as Koyukon and probably English; Neehunilthonoh (Margaret Mayo), daughter of the chief at Nuklukayet who was raised traditionally and would have known her language; or her cousin Seentahna (Jennie Bosco Harper Alexander), who spoke her language most of the time, or some of their relatives, were more likely the sources for Petroff's information about the Tanana names, rather than Paul Kandik.[122]

Upstream on the east bank of the Tanana, the name Saklakageta* on the map is probably the Salcha River, perhaps derived from a Salcha speaker's name for the river, Salchaket. Further upriver on the west bank is a tributary marked Tutluk*, which perhaps is somehow related to the area around Big Delta. For example, the name for the hill beyond Big Delta recorded with Salcha speaker Abraham Luke is Taach'aa Ddhele[123] and perhaps that was attributed as the name for this river by Mercier or another white trader, or by Paul Kandik, or another Native guide or speaker acting as informant.

Altogether these eleven Native place-names on the Yukon and Tanana provide plenty of room for speculation but no definitive information about Paul Kandik's identity. Yukon linguist John Ritter has noted that the different endings on the tributary names along the Yukon, which suggest both Eagle and Klondike speakers as sources, may indicate that Mercier drew on more than one guide when learning local names initially. For whatever reason he recorded a variety of names on the map—either with those different linguistic sources and guides present, or drawing on his memory of the names spoken by many different Native trade contacts he had met since his arrival on the Yukon in 1868. If only one speaker were the source, the names would more likely have been "normalized" to one form of pronunciation and endings, either reflecting an Eagle speaker's version with "juu" as the ending or that of a Klondike person with a strong "dek" sound.[124]

As noted above, one name, the Tchanindou,* appears to have been attached to a different tributary than the one identified by that name today. If the name is on the wrong tributary, this suggests more interesting questions about Mercier's geographical knowledge, and of the mapmaking process. How could this have happened if Mercier and Kandik were working on the map together? Perhaps they were not in the same place at the same time at all, with Kandik drawing his outline and Mercier adding names at a later time and place. The inscription on the map says it was received from "M. Mercier at St. Michael, July 1, 1880,"[125] which was about two weeks before Petroff started traveling up the Yukon with McQuesten on the Alaska Commercial Company steamboat.[126] Mercier was working for the "opposition" Western Fur and Trading Company then and had his own steamer called the *St. Michael* and Native crew.[127] Paul Kandik might have been a pilot or worker on one of these boats, but again no mention is made by name of Native crew in Mercier's memoirs, nor in any of Petroff's reports. Did Petroff ask the traders and their Native associates at Saint Michael for the map to be drawn in advance of his trip upriver? Nothing in his correspondence or his published sources provides more clarification on these questions.

Some tributaries were not named on the map. For example, the name on the next river down from the Kandik on the left bank appears to have been erased. This is likely the Charley River and the erasure of the name gives rise to further speculation about Mercier's knowledge of the Native place-names and perhaps even the geography of the river. In 1874 he made his first steamboat trip up the Yukon as far as Fort Reliance when he left Jack McQuesten there to establish the new post. He may have made the trip upriver a few more times by 1880, but had never lived in that area at that date, although his brother Moïse was at Fort Yukon for several years and would have passed on information to him.[128] The Yukon River is long with many tributaries both large and small—as a relative newcomer to the region, François may simply have confused some names for tributaries and been uncertain of others, or in some cases saw no need to state the obvious. For example, the Porcupine River is not named, nor is the Yukon by that name anywhere along its course. This may indicate that these rivers were so familiar to Kandik, Mercier, and the intended audience for the map as to need no labels. At the time the river above the confluence with the Pelly River was known as the Lewes, and this was the name attached by Mercier, together with a French spelling as Louis R.*

Overall the greatest number of Native place-names on the map is located in Hän country, recording at least two variations of Hän endings for river names, further supporting the idea that Paul Kandik was a Hän man. Kandik's drawing is remarkably accurate for the river courses over this vast region but he may not have known names for all the tributaries beyond his personally traveled and traditional areas, for example on the Tanana and the Kuskokwim. The uncertainties surrounding the naming of some rivers in Hän country may indicate gaps in communication between him and Mercier and perhaps Petroff. It seems likely that Kandik could have been at Saint Michael with the upper river white traders in order to draw the map prior to Petroff's Yukon River trip, but there is no documentary evidence to confirm that idea.

Conclusion

Kandik's river craft and geographical knowledge would have been honed during the days of birchbark canoes. It is certain that his Elders would have passed on Yukon River knowledge accumulated over several generations.[129] H.B.C. trader Alexander Hunter Murray noted that three of the Hän men trading on the Porcupine whom he met in 1847 had traveled previously all the way downriver to trade at the Russian post at Anvik.[130] It was common practice for the Hän people to trade at Nuklukayet at the confluence of the Tanana with archaeological evidence and documentary and oral traditions identifying that site as a location for annual gatherings among many groups of Tanana and Yukon River people.[131] Hän people also traveled upriver to trade with the Wood Indians or Northern Tutchone in the Pelly area and over to the Tanana to exchange goods with the Mountain people.[132] In earlier days indigenous travel and trade was more limited to the territories of a particular group and their immediate trading partners. In the days of H.B.C. and Russian rivalry on the Yukon, some Hän people traded with both companies but did not travel the full length of the Yukon.[133] Seasonal limitations and the need to gather food prevented longer excursions and those conditions prevailed when Raymond and other early travelers failed to persuade Native guides at Fort Yukon to paddle all the way to the Pacific with them.[134] Paul the mapmaker clearly had a long reach to his geographical knowledge, so perhaps he was one of those who made the transition to piloting for the earliest steamboats which arrived for the first time at Fort Yukon when he would have been a young man in his late teens or a little older. More research is needed to determine when upper river people began traveling all the way to the Russian and later American posts at the sea.

Paul Kandik must have been closely associated with nonnative traders and travelers, for his drawing addressed the key questions and interests of the nonnative people of this era—how to travel from the Yukon to the Tanana, what routes to take to the Kuskokwim, which trails led from the Pacific Coast to Fort Selkirk, and where the meeting places and posts of traders were located. These were all points of interest and knowledge essential as a Native paddling guide, and later as a steamboat pilot on these rivers, and perhaps point to possible sources of additional clues to Paul Kandik's identity. More research on early river travel may yield some information, or some report of Native river pilots in the early issues of the *Sitka Alaskan* or Juneau newspapers, or perhaps in the letters and journals of other early visitors.

For the time being the label on the Kandik Map remains the only evidence of a person called Paul Kandik. The map tells us that he had an intensive and extensive knowledge of the territories that were central to the lives of Hän and Tanana people, that he must have been well traveled, to have known, remembered, and drawn with confidence the grand riverscapes of the whole interior of Alaska and Yukon with the overland routes between them. He probably added information gleaned from white men like Mercier and Petroff and their maps that expanded his personal knowledge. Though the evidence on the map is limited and tenuous, it sheds light on his possible origins, linguistic affiliations, and lifestyle. The documentary sources compiled by nonnative contemporaries, including traders and missionaries, provide more insights into Paul Kandik's world and the times in which he lived.

3

Documenting a Mystery

THE LACK of memory about Paul Kandik among Hän people is understandable given the turbulence of these times for Native people. What accounts for the fact that he also remains an enigma in the documentary sources left by his nonnative contemporaries, especially François Mercier and Ivan Petroff, who were both on the scene in 1880 and apparently instrumental in producing the Kandik Map? This absence of Paul Kandik from the records and publications compiled by nonnative writers is typical for the period as Native people generally were referenced only briefly and nonspecifically, identified as "an Indian" or "Native guide" rather than by personal name.[135] The reasons for this practice reflect the prevailing views of many Europeans, and nonnative Americans and Canadians, that indigenous peoples worldwide were inferior races destined to disappear with the arrival of newcomers.[136] There were, as well, fundamental and practical barriers to communication owing to language and cultural differences,[137] some of which still persist today as issues in tracing the story of the map. The takeover of Fort Yukon by Americans in 1869 and the departure of the Hudson's Bay Company from the Yukon River also had a major impact on the continuity of record keeping and subsequent availability of information.

Tracking the Names Paul and Kandik

How did the names Paul and Kandik come to be linked and associated with the man who drew the map? A safe assumption is that Paul Kandik was born sometime between 1850 and 1860, in order to be an experienced river traveler of twenty to thirty years in age, and capable of drawing the full length of the Yukon and related tributaries on his map by 1880. As a Native person of that era he would have been a continuous resident of the Alaska-Yukon region from birth to death, most likely located somewhere along the Yukon or upper Tanana rivers, and probably living in several different locations over his lifetime. If the preponderance of place-names and details of landscape for the section of the Yukon between the Klondike and Porcupine rivers plus other evidence is indicative of his identity as a Hän man, there could be several explanations for how he acquired his names.

During the 1850s when he would have been a child or a very young man, no missionaries had traveled in this region. Most Native people had not yet been baptized with Christian names and were still known by their Native names. A few of the most important chiefs and headmen dealing with the fur traders were given descriptive English names, such as Red Leggings,[138] or an anglicized or francophone version of their Native name, such as the various renderings of Shahnyaati'.[139] When missionaries came to the region in the 1860s and 1870s, more Native people received English first names at baptism, although only a few were given English last names in the early years, and most continued to be recorded with English first names followed by their Native names. By the next generation many people took their father's English first name as a last name and acquired an English first name as well.[140] Paul the mapmaker probably received his English first name at a baptism ceremony sometime in the 1860s or 1870s, although he is not identified in the most comprehensive church records for those years, the journals of Anglican missionary Archdeacon Robert McDonald.[141] Roman Catholic missionaries, who were guests of

Moïse and François Mercier in the mid-1870s, may have baptized him, but he does not appear in any records located to date.[142]

Later nonnative traders and travelers also had difficulty pronouncing Native personal names,[143] and it appears that the people most frequently associating with them, such as chiefs and guides, adopted or were given English personal names and sometimes also place-names to identify them (examples include Dawson Charlie, Tagish Charlie).[144] Perhaps, a trader, missionary, or Petroff attached the name Kandik to Paul in order to distinguish him from other men named Paul along the river. Native people also had a tradition of using places to identify people (Copper Joe, Big Salmon Pat, and many others) so perhaps Paul Kandik was so named by himself or his own people, though they would likely have used the place-name first and called him Kandik Paul.[145] Today the name Paul survives among Hän people and other groups and is used extensively as both a first and last personal name, though Kandik does not.[146] Kandik does persist as a place-name, and appears to be attached to the same river today as the tributary identified with that name on the Kandik Map. As previously noted, Mercier and Kandik may have been the first to record that name for the river, as earlier maps appear to use the name Antoine River for the tributary on this section of the river.[147] Kandik, or some version of it spoken differently in various dialects to designate "Willow Creek," was probably a Native place-name long before Paul the mapmaker and his map were identified with it. As a personal name for Paul, it may have been used only by some of his nonnative associates, or perhaps it was a name "invented" by Petroff to provide a specific identity for the Native person who drew the map.

If the name Kandik indicates that he was born at the Kandik River, then he was probably a Hän man and part of the band identified by nonnative travelers in the 1870s and 1880s as Chief Charley's Indians, who sometimes camped right at the mouth of the Kandik River and at other times were reported to be across the Yukon at the mouth of the Charley River. He might also have been a Hän man from one of the other nearby and closely related groups, David's Band, who lived near present-day Eagle, or Chief Catsah's Band, who sometimes camped across from Fort Reliance and also had a fish camp at Tr'ochëk at the mouth of the Klondike near present-day Dawson. Since the Hän people traveled extensively to various camps during their seasonal hunting and gathering rounds to visit kin and trade upriver and down, and in later years to act as guides or participate in other wage activities, Paul Kandik could have been living at numerous different places throughout his lifetime.[148]

The name Kandik may denote the place where white traders or missionaries first met this Paul, but it does not necessarily mean that he was born there. Depending upon his age he might have been married to a woman of Chief Charley's band and living there with his wife's relatives when white people such as Mercier or Petroff encountered him. Moving to live with his wife's family would have been the normal course of events for a young man in the matrilocal Athabaskan society of that time.[149] In that case he might have been born to close neighboring groups up or down the Yukon or over on the Tanana, as marriage ties between the Hän bands and other groups were one means of solidifying traditional aboriginal trade links. McDonald reported that one of Shahnyaati's sons was married to the Chief's daughter at Charley's Camp.[150] More research with the descendents of these groups may yield additional clues. For example, the Paul family of Eagle may be related to Paul Kandik, although the most recent ethnohistory on the Hän does not make that link.[151]

If Paul was a Hän Hwëch'in, why has neither his name nor knowledge of the map he drew survived in stories handed down to Hän people today? The absence of stories about the map is understandable since the information it documented was a matter of everyday knowledge to the Hän, and the means of passing on details of landscape and river routes within their own communities would have been through oral traditions transmitted from Elders to younger people, not through documents. The reasons for "losing" Paul Kandik as a name in the last one hundred years are more puzzling, but likely rooted in the fast-paced and massive changes that overtook Hän society. Starting in the 1840s with the arrival of a few H.B.C. traders and missionaries, these changes accelerated rapidly with the discovery of gold in paying quantities in the mid-1880s, when hundreds and later tens of thousands of newcomers came into their lives and their lands. The resulting epidemics, changing trade and transportation patterns, new economic opportunities, and challenging ideas about spiritual and societal values disrupted traditional Hän social groupings and displaced many people from their original lands, leading to knowledge gaps between generations and the loss of Native language fluency for most Hän people within a few decades.[152]

The arrival of missionaries in particular brought impetus to changing Native identities, with new Christian names replacing Native personal names starting with Robert McDonald's visits in the mid-1860s. McDonald's journals connect some of the most prominent chiefs' Native names and their new Christian names, but not all of the people he baptized were documented in this way.[153] In the turbulent times of the Klondike Gold Rush, Chief Isaac of the Tr'ondëk Hwëch'in decided to entrust the preservation of his people's most treasured songs and dances to related groups in Alaska because he feared these traditions would be lost in the chaos of Gold Rush society.[154] His concerns proved to be valid as the number of Hän speakers in the Klondike dwindled, along with the knowledge of songs, dances, and other oral traditions such as personal names. Very few of the traditional names in the Hän language remain in oral currency today, except for some of the principal chiefs identified by the newcomers in their records and passed down through ethnohistorical research and recent commemoration. If Paul Kandik's Native name were documented somewhere and linked to his English name, it might be possible to trace him to some of the genealogies constructed in recent years through oral histories by current descendents of Hän families at Dawson and Eagle, or to connect him to Tanana or Gwich'in communities.

Once the Klondike Gold Rush was fully engaged, the international boundary line between Alaska and Yukon became more important to the United States and Canada. The people of earlier days from Charley's Camp and David's Village were drawn into new systems of organization for schools and other administrative structures that increasingly separated people on either side of the international borderline, disrupting family connections and undermining Native languages, muting the transmission of stories and other information through oral traditions.[155] Paul Kandik may have died during one of the numerous devastating epidemics that killed many Hän and other Native people during these years.[156] He may have left his people to work on one of the big new steamboats that replaced the earlier small steamers in the rush of 1897–1898 when he could have been in his forties or fifties, depending upon his age at the time the map was drawn. He may have been caught up in one of the later stampedes and moved on to a place where he lived and died far from his kin. More likely he was simply not known by that name in his own Hän or other Native communities. His may be one of hundreds of names "lost in translation" during these years, when some people made the transition to English Christian names, while many others died among their own people, identified only by their Native names and not recorded in any permanent nonnative documentary source.

First Meetings: The Hudson's Bay Company Period

The earliest documentary records pertaining to the Hän and other nearby Alaska and Yukon Native peoples were compiled by nonnative travelers, traders, missionaries, and government officials. These sources constitute a fairly small number of records for the 1840s through the 1860s: the journals and correspondence of H.B.C. trader and founder of Fort Yukon, Alexander Hunter Murray (1847–1850) and his successors to 1869; the journals and correspondence of Anglican missionaries such as Robert McDonald (starting in 1862); the publications of American explorers and government officials Robert Kennicott (1859–1862), Captain Charles Raymond (1871), William Healey Dall (1870), and English traveler Frederick Whymper (1869), and a few more. Most of these sources have similar limitations in being focused primarily on the concerns of the newcomers—exploring and claiming what was "new" territory for them, surviving in a harsh, unknown environment, and discovering sufficient resources to make a living—rather than detailing the identities of Native people they met.

Many early writers were in the region for very brief periods—sometimes as little as a few days in any one place or for a few weeks at most on a seasonal basis. The few who stayed for several seasons, such as Murray and McDonald, had a more extended opportunity to meet and learn about their Native neighbors, but they too were hampered by language and cultural differences. Most were traveling quickly through the country, meeting Native people all along the way, but rarely staying long enough at any one location to develop a clear understanding of names and family associations. As a result they included very little information concerning specific individuals in the various Native camps. Of the hundreds of Native people reported as visiting Fort Yukon in the 1840s to 1860s, for example, only the chiefs of the local bands—Red Leggings, Shahnyaati', Charley, and Catsah—are identified by name in early published sources.

H.B.C. traders were meticulous record keepers, being resident in the country for successive seasons and dependent for their commercial success on knowing Native fur suppliers and hunters, their preferences for goods, and skills in trapping. Starting with Alexander Hunter Murray in 1847 and continuing with other traders through the 1850s and 1860s, Company officials recorded the names of more than one hundred heads of families in trading account records and post journals for Fort Yukon, along with notes about who were the chiefs and other details. All are listed with Native personal names, with no English or French names, and neither Kandik nor Paul appears in these lists. There are two names which could possibly relate to the name Kandik: "Kay-sah" and "Kay-zuck," both listed for 1850 and 1856, may be names incorporating "kaii," the word for willow in Hän or one of the other languages spoken by groups visiting the fort, and rendered in this form by Murray.[157] Perhaps one or both were from the Kandik (Willow) River, and possibly one of these names may even refer to the man later known as Paul Kandik, but to date no other records or oral traditions have been found to forge any links between them.

These names were all recorded at a time when neither Native people nor newcomers were able to communicate effectively in each other's original language. As a result most conversations were carried on as a series of translations with the aid of interpreters, and written forms varied widely between writers as they each attempted to record what they heard of the Native languages. The process of translation was complicated by the dozen or more Athabaskan languages spoken along the Kuskokwim, Tanana, and upper Yukon, as well as English, French, and Russian among nonnative traders and missionaries, and several forms of trade patois—Russian Creole, Broken Slavee, and Chinook—with many Native and nonnative people learning some of everything to communicate among so many different groups of people.[158]

Robert McDonald described the intricacies of northern communication in his early letters and journals: "In March 1863 I visited the Indians north of Fort Yukon [and] translated the Decalogue into Tukudh the native language for the first time. . . . An interpreter traveled with me on that trip and so I was enabled to translate the Decalogue on that trip. The interpreter was the wife of a French Canadian and spoke French but no English. I spoke French well."[159] The French Canadian husband was the Métis H.B.C. employee Antoine Houle, who was married to a

Gwich'in woman from the Fort Yukon area. In 1865 McDonald recorded that his translators were some of the most influential chiefs of the region: "April 21—accompanied by 10 of the Indians with sleds loaded with meat for the fort, among others Bikeinechatti, Black River Chief; June 1—1866—set out in canoe with a party of Indians who go to trade with the Indians at the confluence of the Tununa with the Youcon. Among the party are Bikeinechetti and Sahnyatti; June 4—Tununa—many Indians—Tununkutchin and Tetsikutchin—assembled . . . talked to them of God and Jesus, aided by Bikeinechetti and Sahnyatti."[160] In November of 1866, McDonald reported that he "preached in tukudh without an interpreter,"[161] but the following spring when he traveled among a new group speaking a different language, he required a translator again: "March 12—met 3 Kitlitkutchin—old man speaks Tukudh and translates to younger men."[162]

The English traveler Whymper witnessed McDonald's preaching during his stay at Fort Yukon in the summer of 1867, and though impressed by the missionary's work, was skeptical about the success of translation when it came to complex religious teachings: "Rev. Mr. M'Donald . . . held several services with the Indians, addressing them sometimes directly, and sometimes through the fort interpreter, Antoine Houle—a man who speaks French, English, and any number of Indian dialects. . . . [McDonald] has taught some of the younger people to read English . . . [but] with . . . half a dozen different tribes, speaking with as many dialects, it must be very questionable whether they all understood the missionary's words. As in other places, so here is a general jargon called 'broken slavee' used. . . ."[163] Two years later Captain Charles Raymond was unable to speak directly to Native people who were curious about his activities at Fort Yukon, but felt he was able to exchange some ideas of mutual interest. He commented that they were some of the "finest Indians" he had ever met. Raymond's report is a useful source on the people he met along the Yukon River as far as Fort Yukon in 1869, but contains no mention of individuals by name, and Paul Kandik would have been too young at that point to be notable to a visiting government official.[164]

Robert McDonald's records are the one source besides H.B.C. records to include comprehensive lists of the Native people he met for these years. As a missionary each individual was important to him as a convert and, after some Christian instruction, as another person to baptize. Despite his rapid progress in learning the Native languages, the journals illustrate his difficulties in hearing and recording the complex combinations of syllables unfamiliar to the ears of newcomers. He often listed three or more renditions of the names of people he visited most frequently: For example, the Black River Chief's name is variously spelled "Bicheinutti," "Bikeinechetti," and "Bikeinechatti," while the famous Shahnyaati' of the Ramparts is listed as "Sahnyatti," "Shaynuti," and "Sahinyati." Only when McDonald started baptizing people with English names in the 1870s did he establish some consistency in spelling the names of his Native acquaintances, not surprising since he bestowed the names of leaders of his sponsoring agency, the Church Missionary Society, such as David Anderson on Bikeinechatti, and of fur traders such as John Hardisty on Shahnyaati', and his brother Kenneth's name on another of his followers.[165]

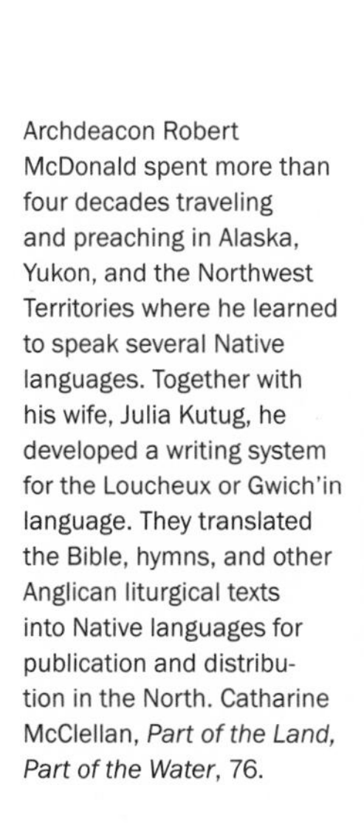

Archdeacon Robert McDonald spent more than four decades traveling and preaching in Alaska, Yukon, and the Northwest Territories where he learned to speak several Native languages. Together with his wife, Julia Kutug, he developed a writing system for the Loucheux or Gwich'in language. They translated the Bible, hymns, and other Anglican liturgical texts into Native languages for publication and distribution in the North. Catharine McClellan, *Part of the Land, Part of the Water*, 76.

McDonald made several trips from Fort Yukon upriver into Hän territory in the years from 1864 to 1890. Of all the early recorders then, McDonald would have been the most likely source to have listed Paul Kandik among his journals and letters. Yet Kandik is not listed in the journals, at least not by that name. He may be one of the many people identified by their Native names in the early journals and somehow missed in later years when McDonald was actively baptizing people with English names on the Yukon. Perhaps by then Paul the mapmaker was away from the Native camps when McDonald visited, traveling with the traders McQuesten, Mercier, or Harper on their early explorations up and down the Yukon and over to the Tanana. In fact only a small proportion of the more than fifteen hundred people listed in McDonald's journals can be connected to people today, as the Native names are rarely cross-referenced to English names, especially for the earliest years of his work around Fort Yukon and vicinity. In comparing McDonald's records to the H.B.C. records[166] a few of the names beginning with K may possibly be connected to the same people, and perhaps as noted previously, their names may be a rendition of an Athabaskan word for willow. The spelling differs, but the name "Katho" in the 1856 H.B.C. records may refer to the same person McDonald lists as "Kyatho" between 1863 and 1866 and "Roderick Kyatho" in 1875, names recorded during his trips to the Fort Yukon area. The H.B.C. name "Kay-sah" in 1850 and 1856 could be the same as McDonald's "Koisa," listed ten times from 1866 through 1891 and also by his English baptismal name Thomas Jay. McDonald does not include a name similar to the H.B.C. "Kay-zuck"; however his hearing and writing of "Koisa" may perhaps be for the same name and person.[167]

At best these samples of names are very tenuous possibilities for connecting Paul Kandik to various archival records for the region, but they do demonstrate the complexities of tracing Native identities during this period. McDonald's journals have the greatest number of names and details for people of this era, but there are many gaps in the records resulting from his prolonged absences owing to his illnesses (1864–1865; 1880–1881), travels to Winnipeg and Europe (1872–1873; 1882–1887), family responsibilities after he married and relocated to Fort McPherson (1876; 1878), and sometimes difficult weather or travel conditions. In addition the American takeover of Fort Yukon in 1869 disrupted the continuity of his records for Fort Yukon, precipitated by his move first to

Rampart House, then to Fort McPherson, and his subsequent shift to brief seasonal travels on the Yukon.[168] If he had resided permanently at Fort Yukon through the 1880s there might have been a definitive reference to Paul the mapmaker with more information about his origins and relationships to past and future generations of Native people.

The American Trade and Exploration Period

The conditions for cultural exchange between Native and nonnative people expanded through the American trade era from the 1870s to the 1890s and through the Klondike Gold Rush and beyond, but with persistent linguistic and cultural factors affecting communications. Athabaskan people continued their subsistence lifestyle, combining the new and expanding trade opportunities with their seasonal hunting, trapping, and gathering rounds. The availability and continuity of primary records for this period is minimal, since it appears the new American traders either did not keep detailed records about their aboriginal business partners or their records have not been preserved.[169] There is a significant number of published reports and manuscripts, but the travelers, missionaries, and government officials of this time continued the earlier pattern of naming only a few of the Athabaskan people they met. McDonald, McGrath, McQuesten, Mercier, Ogilvie, Petroff, Ray, Sim, Turner, and later Canham all contributed significant information about Paul Kandik's world from their perspectives, but no definitive identification for him.

Despite his seventeen-year career as a fur trader in the North, Mercier's memoirs included very few entries pertaining to Native people, and mostly he referred to them as "les sauvages," the common French Canadian term for Native people at the time. Only Shahnyaati', the famous chief so well known to the H.B.C. traders, rated an individual entry in Mercier's writing. He described Shahnyaati' as: "a man of extraordinary stature and strength, of braveness or more than that, ferocity, and duplicity which one rarely encounters even among the savages. . . . Not only the terror . . . of his tribe, but . . . of the other neighboring tribes . . . whom he massacred . . . for the least reason, or for only the love of carnage. . . . I can say in his favor that although very annoying and a beggar towards white people, he never did them one wrong, and often rendered them his services."[170] Mercier mentioned that Shahnyaati's people, the "Koutcha-Koutchin, . . . as well as their neighbors and relatives, the Houn-Koutchin, or Gens des Foux . . . [were all Protestant converts but] . . . this did not stop them . . . from being as superstitious, if not more, than previously. They were then, true hypocrites who would not have wanted to shoot a rifle on Sunday . . . because their minister told them not to, but who meanwhile would steal from you. . . . They were sharp, shrewd men. . . . It was of all the Indians of Alaska, these that I detested the most. . . . They were meanwhile, famous hunters, and many among them were handsome men."[171] Given these views on the upper Yukon people, perhaps it is not surprising that Mercier included no reference to Paul Kandik.

Mercier seems to have had a more positive relationship with other Native people, reporting that he served as godfather to "large numbers" of Native children baptized by Roman Catholic missionaries downriver, and that the son of "a very influential chief of the Upper Tanana" had been baptized with his name.[172] Roman Catholic missionaries attempted to establish a mission at Fort Yukon in the 1860s but received minimal

Shahnyaati' was one of the most powerful Native leaders on the upper Yukon. He traded with the Hudson's Bay Company and later various American companies at Fort Yukon, eventually replacing them all and operating his own business there. He is pictured at the far right standing with two unidentified men and a woman sitting at a table in the background in front of a summer camp, probably at Fort Yukon c. 1883. Alaska State Library, Wickersham Collection, Charles Farciot, SL-P277-017–016.

support from the Anglican H.B.C. factor, McDougall, who favored Robert McDonald, causing the priest to leave after one winter.[173] In the 1870s Mercier used his contacts in Québec to encourage Catholic missionaries to return to the Yukon River, offering them lodging and support at the trading posts under his influence. The Catholic missionaries found the people at Fort Yukon and upriver were steadfastly loyal to their Anglican minister and turned their attentions to groups on the lower Yukon.[174] As a result the Roman Catholic mission records include names of people at Nulato, Anvik, and other places downriver, but not Paul Kandik. This is not surprising since he probably would have visited those communities only briefly as part of a boat crew going up and down the Yukon.

One of the American men Mercier hired in 1874, Leroy Jack McQuesten, provided more extensive descriptions of the important roles Native men were playing during this period when steamboats were introduced and several American companies were vigorously expanding their trade and prospecting territories. McQuesten recounted his trip upriver with Mercier in 1874 to establish Fort Reliance among the Hän people. Mercier was then Agent General of the Saint Michael District for the Alaska Commercial Company and considered this venture to be one of the highlights of his career in the North.[175] The little company steamer was named the *Youkon* and the crew was a mix of Native and nonnative expertise: "Mr. F. Mercier was in charge, Mr. Forbes was Engineer, Mr. T Williams was Pilot, McIntyre was cook, Mayo, Frank Farnfield and myself were passengers. We had an old chief called Catsah and ten of his men aboard—they were Trondiak Indians. As it was the first time the Steamer had been on that part of the River we had considerable trouble in keeping the Channel which necessarily delayed us some."[176] Once landed McQuesten remembered that the Hän helped to construct the new post and were interested and helpful neighbors most of the time. In his account of the founding of Fort Reliance, Mercier made no mention of any Native guides and helpers.

In later life, McQuesten recorded other details of his experiences with Hän people, whom he knew well having lived first across from Catsah's group at Fort Reliance for several years, and then later at Fortymile and Circle City where he would have been in frequent contact with David's and Charley's bands as well. During the winter of 1878–1879 McQuesten wrote about a serious accident at Fort Reliance which might have been fatal without help from his Native neighbors: "In March I fell out of the loft of the store—I struck on a nest of Camp Kettles on my back. I broke one of my short ribs. It was two weeks before I could move and I was in great pain unless I was in a certain position. There were three bands of Indians within days travel—Davids, Charley and the Tronduk—they would send in a messenger every day to hear how I was getting along and the Shoman [*sic*] were making medicine for me to get well and still they were twenty miles away. They thought if I should die that they might be blamed for killing me as there was no other white man in this part of the country."[177]

In 1880, during the summer when the map was drawn, McQuesten and Mercier were working for opposing companies. Both men were operating in Hän country, looking to extend their business, relying on Native men for help in navigating their steamboats, and out of necessity finding ways to join forces on occasion. McQuesten wrote: "we met the St. Michael [with] Frank Mercier . . . in charge, he was in the employ of the opposition. We traveled in company as far as Fort Yukon, we had the best pilots but the poorest steamer and she was liable to break down any time. Mercier proposed to leave our steamer at Fort Yukon and our crew go aboard of his boat and they would tow us up. . . . For the last three years I had been running her after . . . Tanana Station I was the only white man aboard, the crew were all Indians."[178]

McQuesten wrote about his partner Arthur Harper, who traded and prospected up and down the Yukon, often with assistance from Native guides, but there is no specific reference to these people by name. Since Harper traveled extensively in many areas shown on the Kandik Map between the Yukon and the Tanana, it is likely that he knew Paul Kandik, and possible that he was one of the Native men reported to have traveled with Harper. McQuesten described one of those trips in the summer of 1881, probably the first time white traders crossed from the Yukon to the Tanana: "Mr. Bates was aboard of the St. Michael and him and Harper had made arrangements to go cross country to the Tanana River from David Camp. They engaged two Indians to carry their provisions. . . ."[179] Bates sprained his ankle and fell on the north fork of the Fortymile. While Bates dried his clothes, Harper prospected the sands in the vicinity, which Bates saved and sent to San Francisco where the samples assayed at $20,000 to the ton.[180] Next they went on to the Tanana River where they found some people who made them a moose-skin boat and they proceeded on their way down the Tanana to Nuklukayet.[181]

Clearly there was a lot of interaction between Native people and McQuesten, Harper, and Mayo, especially with the Hän, as the traders lived at various times at Fort Reliance, Stewart River, Forty Mile, and Circle City. Photographs of the period depict numerous Native people at their various trading posts. With their Athabaskan wives and children, McQuesten, Harper, and Mayo all had opportunities for extensive connections and communications with Native people, strengthened by kinship ties through their extended families, but none of the women or children or other relatives are named in the traders' accounts.[182] McQuesten named chiefs Catsah and Charley, plus one other messenger named Shernouthy, in his *Recollections*, but none of the boat crews or other Native helpers, and not Paul Kandik.

While Harper left no manuscript or published records, he did write letters to friends and newspapers outside, reporting on the golden opportunities in the Yukon Valley and encouraging other miners to come North.[183] Northern trading companies were advertising and promoting Alaska, as well as lobbying the United States government to support commercial development in its newest territory. Ivan Petroff's appointment to take the first official American census in Alaska in 1880 was one result of these growing American commercial interests in Alaska. He was an enthusiastic chronicler of his trip up the Yukon River, in his various official reports and in correspondence with his wife. Since it is his inscription on the map that links the names Paul and Kandik, his records would be a logical source for further details on the mapmaker, but there are only tantalizing nonspecific references to the Native people he met.

Petroff's writings do provide considerable context for understanding the times in which Paul Kandik lived and the significant Native contributions to his work. Petroff noted that technological changes were transforming the Yukon River, "which has served as the highway of nations and tribes for centuries, long before the white man, with his improved means of transportation, accomplished the feat, marvelous in their eyes, of traversing in one brief season the distance from its deltoid mouth to the Hudson Bay fort."[184] He wrote that "American enterprise has already taken hold of the fur-trade of this region to its full extent, and rival firms have lined the banks of the Yukon with trading stores from Bering Sea to the eastern boundary. The shrill whistle of the steam boat is welcomed annually by thousands along the river banks at the breaking up of the ice, and it is echoed by the hills and mountains of the far interior, where the Hudson Bay Company once reigned supreme."[185] The preface to Petroff's 1882 published report acknowledged the help of "chiefs and other prominent natives of the various settlements during their annual visit to the coast." Significantly, Petroff referenced the maps of Indians and their geographic knowledge as key sources for his 1881 and 1882 report.[186] Paul Kandik's map is the only documentary source found to date to support Petroff's representation of the upper Tanana and Kuskokwim drainages plus the trails between them and the Yukon on the maps included in his reports; earlier American maps such as Dall's and Raymond's left this area blank. Why then does he not mention Kandik by name in his published reports and maps, when he so clearly and specifically identified Paul on the manuscript map? Most likely he followed the conventions of his time in leaving his Indian informants unnamed, though at least he acknowledged their contributions as significant to his work.

In a letter to his wife from Saint Michael on July 8, 1880, Petroff wrote that he was impatient to be on his way up the Yukon, but he had to wait for the traders to complete their business on the coast as he was to travel initially on their steamboat, apparently with some reluctance. "All the traders are pretty rough—not much above Indians in their life and manners. I shall be well satisfied when I leave them and travel with the natives alone."[187] He also wrote: "I have a little Indian deer-skin dress ornamented with beads such as they wear on the Upper Yukon, just about big enough for Olga [his daughter]—or rather two, one male and one female, but there is not much difference between them."[188] This raises an interesting question—did some of the traders bring clothing belonging to their children down to Saint Michael, or did their Native families accompany them on this occasion and trade the outfits to Petroff? Did he acquire the outfits from other Yukon Native people who traveled downriver as pilots and crewmembers on the traders' boats or on their own? He noted that a Tanana man and his wife made the trip downriver that summer—the first to do so, and he included an illustration of them in his report published in 1882.[189] All of these small bits of information lend support to the idea that Paul Kandik could have been at Saint Michael that summer, in order to draw the map by July 1, before Petroff commenced his journey upriver.

Petroff seems to have appreciated his Native assistants, indeed preferring their company to the traders. Still, he did not provide the names of the Native boatmen, who paddled his canoe upriver to Nuklukayet after he left the traders' steamboat at Anvik. Some of his comments were typical for nonnative writers of the time, but it is clear he felt well served by them. In a letter sent to his wife on his return trip downriver, he described the trip as "almost one continuous agony owing to mosquitoes and small black flies. . . . They all bite furiously and I am always bleeding—face, neck, and hands. . . . My two Indians are very good men, they take good care of me and will not let me do a thing either in camp or in the canoe. They have my tent and mosquito bar up as soon as we stop and when they get through cooking, etc. they sit near me and smoke for me—they use horrible tobacco that makes a fearful smell, but I would rather suffer through the nose than be bitten all over. One of them brings me a basin of water to wash in and then both stand over me with smoking firebrands while I perform my ablutions. The same men will remain with me until I reach the seacoast again."[190]

Could Petroff's paddlers have included Paul Kandik on the trip to Nuklukayet? He would have known the route, and if he were introduced to Petroff as a knowledgeable mapmaker he would have been a very desirable

This group of men was identified in the Schieffelin Album as a group of "Upper Yukon Indians" and "Esquimaux" who would have lived on the lower river. Both groups served as pilots and crew in their traditional areas on the steamboats operated by the white traders such as Mayo, McQuesten, and Mercier. This photograph was taken at Saint Michael c. 1882–1883. Alaska State Library, Wickersham Collection, Charles Farciot, SL-P277-017—007.

guide for the long trip. More likely Petroff hired some experienced paddlers from around Anvik, which he observed was an unofficial border between the traditional territory of coastal Yupik people and interior Athabaskans: "At the present time the Indian or 'Ingalik' tribes hold full sway over the river down to Paimute village, situated below the junction of the Anvik River with the Yukon, and no Innuit (or Eskimo) ascends the river beyond this point unaccompanied by white men, while no Ingalit descends without the same protection."[191] Given the other "map" evidence of Kandik's likely origins in Hän country, it seems more probable that Paul would have been one of McQuesten's or Mercier's Native crewmen returning home in July "under steam" to the upper Yukon, rather than paddling Petroff's canoe hundreds of miles up and down the river, ending up at the lower end far from home in late August with ice soon running in the river.

The next official traveler to journey down the Yukon and provide a detailed record of places and people was U.S. Army Lieutenant Frederick Schwatka, who conducted a military reconnaissance in 1883, with particular instructions to provide information on the Native people of the region. Schwatka hired Tlingit guides (Chilkoots) and some Interior Stick Indians (Tahkheesh) camped on the coast to pack his heavy loads across the Chilkoot Pass, the majority packing to the headwaters of the Yukon where Schwatka paid them off before they returned to the coast. Some of the Tlingits carried on with the party, acting as interpreters and providing information about their routes to the interior, plus the names of rivers and camps of the Stick Indians.[192] Two of the Coast Indian men continued on down the Yukon with the party: "one was a half-breed Tlingit interpreter, Billy Dickinson [who] understood the Tlingit language. . . . Our other native . . . Indianne, a Chilkat Tahk-heesh Indian, whose familiarity with the latter language, through his mother . . . made him invaluable to us as an interpreter while in the country of this tribe, which stretches to the site of old Fort Selkirk at the mouth of the Pelly River." Indianne was also valuable as a guide "having traveled over parts of it much oftener than most Indians, owing to the demand for his services among the Sticks. Through the medium of our two interpreters, and the knowledge found in each tribe of the languages of their neighbors, we managed to get along on the river until English and Russian were again encountered, although we occasionally had to use four or five interpreters at once."[193]

Schwatka's interpreters were less familiar with the people and the languages as the party passed through Hän country in mid-July, but he and John Wilson, the surgeon for the expedition, recorded some names and descriptions of the camps and circumstances of the Hän they met. The first camp of "Chief Chil-tah" (probably Catsah) they understood to be called "Noo-klahk-o," and noted that the people there had a plentiful supply of goods obtained from the Alaska Commercial Company trader at nearby Fort Reliance, which was supplied by a steamboat annually. They still obtained some items such as ammunition from Chilkat Indians and some goods from the H.B.C. post on the Porcupine. Native Christian leaders were teaching the people new beliefs, but they were also still under the influence of traditional shamans, of whom two were named and briefly described by Schwatka. "Ee-nuk" was said to have a very bad temper, and the other was named "Joseph." The people called themselves "Takon . . . as well as could be determined through our interpreter."[194] Schwatka counted twelve houses here and later heard from trader McQuesten that eighty to eighty-five people belonged to the group, but many more were present when Schwatka was there in July as there were visitors from the Tanana. He met some Native people just above Fort Reliance from whom he learned that the white trader from that post (McQuesten) had gone downriver.

At Johnny's Village or Klat-ol-klin[195] (Mercier's Thetawdé and today's Eagle), Schwatka's group met more Hän people and learned of another white trader, whom he called Mercer (Mercier), and his nearby Belle Isle post. These people were called the "Klat-ol-klin Indians . . . by a trader's half-breed Russian interpreter who has lived among them for several years," but Schwatka's interpreter learned they designated themselves as "Takon" like the people at the previous camp, and called their camp "Johnnys Village" after their head chief who was away at the time. Here Schwatka noted that the "custom of shaking hands copied from whites, means shaking hands with everyone in the village."[196] Further downriver Schwatka visited the "Tadoosh or Charley's Indians" with fifty to sixty people living in five or six "houses of brush," all related to the people at "John's Village" and Fort Reliance. This group was reported to be friendly towards whites by the white miner (Joe Ladue) who was there waiting for the arrival of the steamboat. Charley was the chief and the only person named at this location.[197]

A few days later Schwatka landed at Fort Yukon, noting that it had been abandoned by white traders and occupied by Chief Shahnyaati', who was given goods by them to trade for furs to other Native people. There the party also encountered the Alaska Commercial Company steamer *Yukon* on its annual trip upriver, and later the *New Racket*, which had just been purchased by Harper, McQuesten, and Mayo operating as independent traders.[198] Schwatka did not mention any Native crew on the boats, though it is likely he saw them. At Nuklukayet he met Arthur Harper and learned that this post was functioning as the "frontier trading station on the Yukon," with all the posts upriver being abandoned for the time being as a result of "strained feelings" between the traders and Native people after fur prices fell with a recent amalgamation of the American fur-trading companies. Harper said that the Tanana people who came to trade at the post were friendly, despite a reputation for being warlike and uncivilized, and recently had asked for a missionary to come and teach them. Two of their chiefs were named "Ee-van and Jack."[199]

Lieutenant Frederick Schwatka visited the people camped at Tthee t'äwdlenn/Eagle Bluff near Mission Creek, which he called "Johnny's Village or Klat-ol-klin," also known as David's Village and now as Eagle Village, on his voyage down the Yukon in 1883. His topographer produced lithographs for publication with Schwatka's official report, which were adapted for his popular travelogue on the trip. Pictured here are the birchbark canoes, fish drying racks, and homes of this Hän village. Frederick Schwatka, *Along Alaska's Great River*.

Schwatka was traveling through Hän country just three years after the Kandik Map was drawn and certainly met people who would have known Paul Kandik, if he did not meet the man himself. He might have had access to Petroff's preliminary census reports of 1881 and 1882 in the United States before he departed on his expedition, together with the maps incorporating Kandik's information on Yukon tributaries. Schwatka's military orders were to assess any threats represented by Native tribes in interior Alaska, and he concluded that none of the groups he encountered posed any danger; in fact, the people were friendly and very interested in meeting his party. Native men working as guides, pilots, and boat crew probably would not have appeared threatening and would have been busy working or resting when Schwatka met the traders in their steamboats. This could

Ivan was one of the leading men of the Tanana people trading at Nuklukayet during the 1880s who impressed Allen, Farciot, Schwatka, and numerous other early travelers in the region. A handwritten note on this photograph says he was "A great Indian orator." Alaska State Library, Wickersham Collection, Charles Farciot, SL-P277-017–013.

account for them not being named, while shamans who were ill-tempered and possible instigators of trouble for nonnative visitors were more remarkable in Schwatka's eyes, and so named in his report. Schwatka lacked interpreters who could communicate with local people once he traveled below the White River into Hän country, which could account for his assertion that the people at "Tthee t'äwdlenn" (Eagle) were called "Johnny's Camp" after a local chief when other observers consistently named this group "David's Band."

Two years later, in 1885, Lieutenant Henry Allen led his American Army expedition north from the coast at Prince William Sound, up the Copper River, over to the Tanana and down its entire length to Nuklukayet, making the first official survey maps of this region. Allen had a lot of help from numerous Native people who fed his party, told stories of their travels to the Yukon, drew maps, and warned of various dangers along the way. "I had frequent maps made by the natives to show us the trail over the Alaskan Mountains and down the Tanana to the Yukon River." Allen traveled with Chief Nicolai, who recalled the route to Nuchek and helped Allen sketch it in with a dotted line on his map. Another Tanana man drew a map of the Yukon and the Tanana, which Allen included in his report "to show how great is the geographical knowledge of these primitive people. He assured me he had been to the stations on the Yukon, at Fort Reliance and at Fetutlin, the former kept by Mr. McQuisten [*sic*], the latter by Mr. Harper. . . . He was entirely ignorant of their surnames but spoke of 'Jock.'" Allen met people on both sides of the Copper-Tanana divide who had been to the Yukon to visit, stories confirmed at Nuklukayet where he met "McQuisten [*sic*] the trader at Fort Reliance and La Due [*sic*], a prospector, who said they had seen some copper Indians . . . in 1883. . . [and] . . . that a native on the north side of the mountains was used as a second interpreter to them."[200] Could that interpreter have been Paul Kandik? Certainly the information given to Allen along the Tanana confirms the many other reports of travel and trade between Hän and Tanana people, along with the knowledge of trails and the ability of numerous Native people to draw detailed maps of the area.

Allen met many helpful and informative people on his travels, but named very few of them. He described one old man who mapped out the rivers, tributaries, and portages of the Koyukon region. Allen at first doubted his knowledge since he counted the days of walking on his toes but later had to admit the man was proven correct on several counts. He met "Red Shirt," a "Koyukuk chief" who had guided another American officer, Lieutenant Cantwell, and noted he was implicated in the Nulato massacre of the 1850s. He also met "Manook," whom he described as "the interpreter at Fort Reliance," who spoke all the dialects and was considered one of the best interpreters along the Yukon but said he had never been on the Tanana.[201] In completing his report and maps, Allen acknowledged the geographical contributions of many earlier nonnative travelers in the region—Dall, Raymond, Schwatka, Petroff, and others—but if he was aware of any of their Native informants, such as Paul Kandik, he did not name them.

Anglican missionaries were still active on the Yukon in this period, with Robert McDonald, his brother Kenneth, and Vincent Sim making trips down the Porcupine and the Yukon to the Tanana, and up as far as Fort Reliance in the 1880s, but none of them mentions Paul Kandik either. Robert McDonald does mention a man named Paul, whom he met at Harper's trading post at the mouth of the Stewart River in 1887. The journal entry is brief: "July 18. At 11 AM came to trading post at mouth of Stewart River. Had divine service with an Indian and his wife there encamped. The woman and five children all baptized. The man Paul candidate for Baptism. Three miners here."[202] McDonald continued up the Stewart to visit more miners and "Tutchuntetkwichin" along the river, including "Chief Hanyin." On July 27 he returned to Harper's post and baptized Paul. Could this be Paul the mapmaker, now settled near the post established by Harper in 1884 and perhaps remaining there after Harper moved to Forty Mile in 1886? McDonald provides no further details to identify this Paul and his family.

The following year Canadian surveyor William Ogilvie arrived in Hän country as part of the Canadian government's Yukon Expedition. His assignment was to locate and mark the boundary between the United States and Canada on the Yukon River, a matter of some concern with growing mining interests in the region and the publication of Schwatka's report, which clearly demonstrated American interest in the region, including travel through Canadian territory with no notice to Ottawa. Ogilvie's report and accompanying map published in 1888 included considerable detail about Hän people in the boundary region but no mention by name of Paul Kandik.[203] The diaries of his nephew Frank Sparks, who accompanied the party as an assistant, contain only a few references to Native people in the area and no mention of Paul Kandik.[204]

Big Paul: Pilot on the Upper Yukon

The next year a U.S. Coast and Geodetic Survey party under the leadership of James McGrath was sent to reexamine Ogilvie's calculations and location of the boundary line on the Yukon. Arriving in the fall of 1889 the American surveyors took up residence in the cabin Ogilvie's party built in 1887.[205] McGrath completed detailed daily journals and monthly reports of his party's activities over the course of the two years they spent here from August 1889 to spring 1891. The records of the first season are available at the U.S. National Archives II in Maryland, but those for the second year are missing from the bound volumes of reports. In the reports for 1889–1890 McGrath mentions the local Native people occasionally, naming Chief Charley, who was a frequent visitor at the camp, and Esau, who was the assistant to the Anglican missionary Ellington at Forty Mile.[206] If the missing volumes for the second season could be located they might yield more valuable information as the party developed friendships and had frequent contacts with Hän people during their second year at the camp.

Fortunately the medical doctor attached to the party, Dr. Willis V. Kingsbury, also kept a daily diary, recently transcribed by his great-grandson. Kingsbury wrote lively descriptions of the Native and nonnative people who visited the surveyors' camp and of their nearby neighbors at Forty Mile, whom he visited on several occasions.[207] Kingsbury had a camera of his own, and the party also had a government camera, provided by George Davidson in San Francisco before they departed for the North. The surveyors named their northern home Camp Davidson in his honor, and upon their return they presented him with a commemorative album of photographs documenting their adventures on the Yukon.[208] The album is a remarkable collection of forty-eight images of the people and places McGrath's party encountered from Saint Michael to Forty Mile, including eight photographs taken of Native people identified as "A Band of David's Indians who camped about 3 ½ miles above Camp Davidson during the winter of 1890–91" and another labeled "Indian Paul and son. Paul is one of the pilots of the Upper Yukon river."[209]

Kingsbury was the photographer for most of these photos and his diary documents the names of many of the people and the circumstances when he took the pictures. Most importantly he wrote on several occasions about meeting the Paul he photographed, known as "Big Paul" at the Forty Mile trading post, and whom he first encountered on the steamer *Yukon* in July 1889. While on the boat Kingsbury recorded comments typical for his times, often making fun of the Indian people he met while observing the essential work they did throughout the journey. At Anvik: "One of our men tried to teach some of the native boat hands how to box. Their endeavors were very funny." At Nulato the crew "wooded up" and a shaman came on board as pilot. The survey party was rapidly acquainted with the social import of a steamer's arrival as all the people at Nuklukayet came out to see the passengers. Upriver Captain Peterson shot a beaver and the Native crew skinned it, as well as a porcupine, both later served as dinner to a squeamish Dr. Kingsbury. At the Porcupine River another

Steamer *Yukon* towing three lighters bound for the upper Yukon and Porcupine rivers with the McGrath and Turner U.S. boundary survey parties on board in 1889. Note the photographer at work on the upper deck. Dr. Willis Kingsbury, physician for the McGrath party, had a camera on the voyage and likely took this photograph. The Bancroft Library, Davidson Collection, Photograph Albums 1946.006 Alb, #87.

American boundary survey crew under John Turner was taken up that river on the steamer while the McGrath party waited at Fort Yukon. Turner forgot one of his instruments and an Indian was sent by canoe to catch up to the steamer, returning with a note saying Turner had received it safely. Kingsbury met an Indian at Fort Yukon with a prayer book and skeptically noted that he thought the man only "pretends to read." As they journeyed further upriver the party met more Indian people who could speak English, and who sang hymns as well, the proud students of McDonald and his fellow missionaries. The steamer stopped at George's, Charley's, and David's

Dr. Willis Kingsbury lived at Camp Davidson from the fall of 1889 to the spring of 1891 as physician to the McGrath boundary survey party. He documented his experiences in a daily journal and through a remarkable series of photographs of the people and places he visited. The Bancroft Library, Davidson Collection, Photograph Albums 1946.006 Alb, #40.

camps for wood and meat, finally delivering the party at Ogilvie's camp on August 19.[210]

Within days the newcomers were visited by many of the region's notables from Forty Mile trading post and further upriver, including the Anglican missionary Ellington; trader Tom O'Brien, who was looking for medical treatment; and Arthur Harper with his family from Fort Selkirk. Captain Peterson brought some dogs to McGrath's camp on his return trip downriver. A miner stopped briefly with a Native man as they were going to mine coal on the lower river. Two Indians brought deer and mountain sheep meat, then slept for the night in the camp kitchen. Other Indians came to trade snowshoes, dried fish, and more meat for tea, tobacco, thread, and ammunition. Trader McQuesten sent "a large lot of mittens and moccasins" for the men when the weather cooled through the end of September. More Indians arrived almost daily as news of the survey party and its medical doctor spread up and down the river. Jonathan came for medicine, Jimmy brought more fish, and the chief named Old

Native men were essential to the new economy evolving on the upper Yukon in the 1880s. Here a group of men cut wood near Fort Yukon to fuel the steamboats operated by various trading companies. Alaska State Library, Wickersham Collection, Charles Farciot, SL-P277-017—017.

David camped with his family across the river, becoming regular visitors. Kingsbury was beginning to warm to his new acquaintances, noting that David "is a fine old cuss. Traded a pair of snowshoes for some tobacco, drilling and sugar. He was suffering with a severe cold and [I] gave him a dose."[211]

McGrath's party must have realized the extent of their dependence on the local hunters for food, especially as the steamer *Arctic* had not yet arrived with the bulk of their winter supplies. Kingsbury's diary records ever more anxious listening for the steamboat's whistle day after day as late September brought the threat of ice closing the river. McQuesten sent a note down to McGrath with some Indians to say that if the steamer did not arrive by October 10 he would take all the white miners downriver in his boat, as there were not enough rations in the country to feed them. Kingsbury's entry that day was worrisome: "If the Arctic does not get here we will be awful short on flour etc. McGrath worries about that a great deal." Old David and other Indians continued to bring meat for the camp, which they traded for various goods and medicines. Kingsbury recorded

McGrath's U.S. survey party occupied and enlarged Boundary Hall, the cabin built by Ogilvie's Canadian survey party in 1887. Renamed Camp Davidson in honor of American scientist George Davidson, it was photographed in April 1890 by Dr. Willis Kingsbury. The Bancroft Library, Davidson Collection, Photograph Albums 1946.006 Alb, #25.

that he made "400% profit" on those sales. When Chief Charley arrived looking for payment for the dogs delivered earlier by Captain Peterson, he brought a note from McQuesten advising McGrath to pay him $15 for both. Charley objected saying he wanted $25 but after "a good deal of talk, traded for some of it and took cash for the balance. He wanted some medical advice but when he found out he would have to pay cash to the tune of $5.00 for it he concluded to wait until spring." On another occasion Kingsbury recorded that McGrath "made some astonishing bargains. He paid the Indians 92 cents worth of tobacco and tea for two days work on the river." They had taken two members of the party downriver to catch up with the steamer to retrieve some goods cached at another post. The surveyors were to learn to their chagrin that their Indian neighbors were coming to some interesting conclusions about the treatment they received from the newcomers.[212]

On October 11 some Indians from downriver arrived with news that the *Arctic* had been wrecked and there would be no more provisions

The wreck of the steamer *Arctic* in the summer of 1889, with the loss of all the flour, sugar, tea, candles, ammunition, and other goods destined for mining camps on the upper Yukon, forced a mass exodus of men who could not survive the winter without these supplies. Several dozen miners arrived at Camp Davidson in October from the Fortymile River to commandeer the surveyors' boats for a desperate rush downriver before freeze-up, hoping to reach Saint Michael and the last ships of the season leaving for the south. The Bancroft Library, Davidson Collection, Photograph Albums 1946.006 Alb, #10.

arriving for the winter from the outside. McGrath gave his men the opportunity to leave on McQuesten's boat with the miners for Saint Michael but they all decided to stay for the winter. As government officials they had comfortable rations, including coal oil, candles, and money to pay for local food, which continued to arrive almost on a daily basis: "This evening McQuestions [*sic*] sent us 940 lbs. of meat & two sacks of turnips. . . . I receive two notes from Mr. Ellington describing some sick Indians & requesting medicine for them, which I sent. As one of the Indians that was here (having brought the notes) had meat to pay for it. This gives us nearly 1100 lbs. of meat & will run the camp for about 3 months." Leading up to Thanksgiving the doctor wrote that the camp was on short rations for flour and similar goods but "we have all the meat you can eat, boiled turnips, & some bread. Once or twice a week we have some canned fruit & we have [it] for breakfast over weak oatmeal, cracked wheat and rice. With that layout we pull through and all be fat & in good condition next spring."[213]

With the dark and cold of winter settling around Camp Davidson the holiday was a time for reflection: "Today we got our photograph of our friends who are absent & hung them up around the room. This makes the room a little more cheerful." More good cheer arrived from Forty Mile as Jack McQuesten sent down a bundle of newspapers from outside, letters, and a fur cap for Kingsbury with some Indians who stayed to visit and trade. A few days later the barking of the camp dogs signaled the arrival of a large group of Indians traveling by dogsleds. They brought letters from Forty Mile, moccasins, and moose skins for dog harnesses. Old Charlie "the big chief of the upper Yukon [arrived next] . . . on his way up to visit some of his outlying provinces to hold councils of state . . . followed by his faithful follower Lieut Chief David, who could not deprive us of his company and stayed all night . . . plus 5 more Indians . . . the finest looking set we have seen yet." The Forty Mile traders invited the surveyors to come up to the post for Christmas but they declined, planning to celebrate in style on their own. January brought intense cold with temperatures down to fifty-six below and no visitors for several weeks. Kingsbury noted that the men were looking for some Indians to come in to trade which "would be quite a novelty at present." Finally Ellington sent an Indian with a letter and news of their comrades on the Porcupine. McGrath decided Kingsbury should go up to Fortymile and try to send letters to the Turner party with some Indians.[214]

On March 27 Kingsbury and another member of the party left with the dogs, a sled, and their snowshoes for Forty Mile, taking two days to make the trip. En route they met Walter, an Indian man traveling down to David's Camp, and he drew a map in the snow of the route upriver, estimating how long it would take them to get to the trading post. His drawing helped them reach their destination, but they were completely exhausted with swollen feet, blisters, and wet footwear. Kingsbury was grateful for a warm welcome from the traders, and encouraged about sending news to Turner: "Mr. McQuestion [*sic*] thinks he can get us an Indian to take letters to the Porcupine river party, but he will have to look around amongst them to get a good one & one he can depend upon to get there & back before we are ready to leave." Kingsbury began to learn about the dynamics of bargaining for services in the North: "I had a talk with an Indian, with Mr. McQuestion & wife, & interpreters about going up to the Porcupine. The Indian says one Indian would not go alone on such a long trip as they are afraid of getting sick or cut on the way. Mr. McQuestion says that . . . [on] a long trip like that two or more go & they expect to be shod & fed on the way. . . . Three pair of moccasins will be sufficient for that trip. . . . $20.00, shoes & feed would be as much as should be paid for that trip & he thinks he can get Indians to go for that. This Indian went out to think about it & forgot to come back."[215]

Jack McQuesten stands in the doorway of his trading post at Forty Mile with other residents, some of them his partner Al Mayo's children, and several dogs. The Bancroft Library, Davidson Collection, Photograph Albums 1946.006 Alb, #37.

Kingsbury was still finding it very painful to walk three days later, which may have provided some perspective on the price to be paid for a long and difficult winter trip. There were more serious impediments to making a deal when the next group of hunters arrived to discuss travel: "Big Paul says they are not satisfied with the way they have been treated at our camp & they will not go for our party unless they are well paid for the trip. He says two of them will go up for $40.00 a piece & to be shod & fed. McQ says this is too much & he would not pay it. Paul was very sassy. He wanted pay for a pair of moccasins he gave when we were on the boat. I was treating one of his children & one of the Indians on the boat told me that Paul wanted to give me a pair of moccasins, which he did & never said a word about pay at the time. I would have paid him but McQ said not to."[216]

Kingsbury witnessed another incident at the post which provided more insight into the status of Big Paul and the dynamics among the Indians living and working there: "Tonight Paul took two of the minister's dogs & 4 from an Indian who was hauling logs for the mission house & went back hunting. The preacher & Old David had some hot words about his (Ellington's) dogs. It seems that David wanted to borrow them but that Paul got in ahead of him & got the dogs. This made David very angry & he was going to raise H--- with the preacher but Paul made David go away and behave himself." Apparently Big Paul held sway within his own community as well as being equal to the machinations of the nonnative traders.[217]

Despite the discontent of some of the Indians, Kingsbury recorded more friendly visits at Camp Davidson during the spring. A large group of Charley's band stopped on their way to Forty Mile for their annual caribou hunt: "15 men, 10 women & about 20 children with their camp outfit on 10 toboggans & sleds . . . [arrived] about 9 o'clock this morning & stayed until nearly 5 this evening. . . . smoking their pipes & talking to one another. Sam & one other were the only ones who could talk very much English." Sam declined McGrath's offer to go to Rampart House for $20, saying that "he can make more money for the time working for the miners for $2.00 a day." Kingsbury played checkers with Sam and Charley: "Charley is a good checker player. He did up both Dierks & myself & did it in good style, too. I took pictures of their caravan & a full length picture of Charley." Shortly afterwards another Indian named Little Paul arrived to pick up meat from Chief David's cache and take it back to Forty Mile for him, but Kingsbury provided no other details of this Paul.[218]

Chief Charley was a frequent and welcome visitor at the American surveyors' Camp Davidson home. Dr. Kingsbury photographed him in the spring of 1890 in his caribou skin leggings, mittens, fur hat, and other winter clothing made of cloth. The Bancroft Library, Davidson Collection, Photograph Albums 1946.006 Alb, #30.

These toboggans belonged to some Hän people of Chief Charley's band who stopped to visit at Camp Davidson in the spring of 1890. The toboggans carried all the gear essential for survival on the land, including poles and caribou skins for shelter, snowshoes, and packs for food. Behind the sleds is a dip net for fishing and a rifle plus snowshoes on a willow tripod. The Bancroft Library, Davidson Collection, Photograph Albums 1946.006 Alb, #7.

On May 5 the doctor noted with special emphasis that the "River is open in front of camp today." The group eagerly anticipated receiving news from outside, which came a month later on the *New Racket*, with Captain Mayo bringing sad reports of many deaths that winter along the Yukon. The men at Camp Davidson had survived with the help and friendship of many Native and nonnative neighbors, and Kingsbury's comments were reflective of a deepening respect for the hardy northern river people of all races: "The first mate of the Str Yukon and the engineer of Str St. Michael's have both died. Waschka, the mate . . . was a very good Indian & was a splendid steamboat man now. He had been on steamboats ever since they had been running on this river. We were all very acquainted with him & regret very much to hear of this death. The engineer of the St. Michael's was a brother-in-law of Peterson's & was considered to be one of the best pilots on the river."[219]

In June Big Paul visited Camp Davidson, bringing letters from McQuesten, whose wife was ill and needed medicine. Past disagreements apparently had waned as Kingsbury took pictures of Paul with his son Peter, and one of "Indian Henry." Paul shot a beautiful duck for Kingsbury. They

Breakup was eagerly anticipated by the surveyors at Camp Davidson in the spring of 1890. Joy quickly turned to alarm as the river rose rapidly with enormous chunks coming up over the bank and threatening to destroy their cabin. The Bancroft Library, Davidson Collection, Photograph Albums 1946.006 Alb, #38

saw a boat coming downriver but Kingsbury's hopes were dashed: "this is h—l . . . we have been without mail for a year now & very little prospect of getting anything later than Sept 1899, before next August." The next day was the anniversary of the party's departure from San Francisco and suddenly the men were "startled by hearing a boat's whistle call & all rushed out to the bank & saw the Str Arctic just coming around the bend. . . . Capt Peterson was in command, Harrison engineer, & Indian Louis pilot. There were about 60 miners aboard headed for 40 Mile. They left the mouth of the river on May 31st in reaching this place today they broke the record for steam boating on the Yukon half into it. . . . We got our first mail from the outside today." Among the passengers was the Anglican missionary from Nuklukayet, Reverend Canham, whom the miners and traders didn't like, according to Kingsbury, probably because he was critical of the growing problems with alcohol in the little communities.[220]

Owing to poor observation conditions during the previous winter McGrath had been unable to complete the calculations to confirm the location of the boundary. He decided to stay for a second year on the Yukon, leaving his crew feeling cranky at the prospect of another northern winter. In July two miners and two unnamed Indians came in from Forty Mile on their way to prospect on the Kuskokwim and Cook's Inlet. Little Paul paid another visit, along with several others, including William, who had copper bullets for the doctor acquired from the Tanana Indians. Finally in mid-August the *Arctic* arrived with more outside mail and sad news for Kingsbury that his father had died the previous December. As the men prepared for winter again David's Indians moved to a camp right across from Camp Davidson, planning to stay close by the surveyors, as hunters and neighbors.[221] Kingsbury took a series of photos of "David's Indians" at this camp and one of all his colleagues entitled "Pirates of the Upper Yukon." These photos show two brush shelters with different groups of Hän people posed beside each one, and the surveyors beside a meat cache. In one of the photos two of the men and one of the boys resemble "Indian Henry," "Indian Paul," and Paul's son Peter, with similar facial features, haircuts, and hats as seen in the Camp Davidson photos of the previous spring, meaning that Henry and Paul either were visiting relatives at David's camp or lived there at least part of the time.[222]

During this second winter the surveyors were more sociable with all their neighbors, inviting David's band of men, women, and children to

In June 1890 Big Paul came to visit at Camp Davidson with his young son Peter and Dr. Kingsbury took this photograph of them. Big Paul wore a river pilot's hat while Peter had a distinctive beaded hat similar to those in a later photo identified as "David's Indians" that may include this father and son. The Bancroft Library, Davidson Collection, Photograph Albums 1946.006 Alb, #33.

Dr. Kingsbury took a photograph of Indian Henry in the same location at Camp Davidson on that day. Henry wore a cloth hat here and in a later photo of "David's Indians." The Bancroft Library, Davidson Collection, Photograph Albums 1946.006 Alb, #6.

come for a Christmas "potlatch." McGrath gave them presents of tobacco, tea, sugar, bread, pork, and crackers. Kingsbury concluded, "taken on a whole this Christmas was much merrier than the last one was." One of the young Hän women was particularly attractive but he ruefully concluded that McGrath would not permit her to stay with him in camp. He spent many happy hours coasting on sleds down the riverbanks at Camp Davidson and at David's Camp, where the hill was 200 or 250 yards long: "They use toboggans and sleds. Men, women, and children all take their turns but the women usually have to pull the sleds back. . . . We coasted until dark." He made another trip to Forty Mile and took pictures of the traders and their families. In the spring of 1891 Kingsbury recorded another visit by Big Paul, who had come to visit David's Camp, but gave no more details about him. Finally, on June 22 Kingsbury was on board the steamer *Yukon* again, this time going home after two years of adventure on the Yukon River.[223]

Dr. Kingsbury has left the most complete record found to date of the social life on the upper Yukon ten years after the Kandik Map was drawn, including his photos and many intriguing details documenting the intertwined lives of the Native and nonnative inhabitants, many of whom he identified by name. Of significance to the Kandik Map are the men identified as Big Paul and Little Paul. Could one of them be Paul the mapmaker? Could one or both have moved from Charley's camp at the Kandik River upriver to the Fortymile post after the discovery of gold there in 1886, drawn by a burgeoning new mining economy with lots of employment opportunities, ready access to trade goods, a fascinating new social milieu, and longtime trading associates McQuesten, Harper, and Mayo with their Indian wives and children? It is at least a possibility that one or both Pauls were part of Catsah's "Trondiak" group when he guided Mercier to Fort Reliance in 1874. They would have been well acquainted with McQuesten, having associated with him for over a decade, and possibly worked as crew for Mercier on his steamboat as well. Big Paul is the most likely candidate to be the mapmaker since Kingsbury identified him as a pilot and reported his presence on the *Yukon* as crew in 1889. He was a large man as evidenced in Kingsbury's photo of him at Camp Davidson and obviously had great standing in his Native community to be able to dissuade Old Chief David from causing problems over the dog issue that Kingsbury recounted in 1890. Kingsbury's diary clearly documents the

Dr. Kingsbury visited the people he identified as "David's Indians" in the fall of 1890 when they camped a few miles upriver from Camp Davidson. In this group the man looking directly at the camera may be Big Paul in his pilot's cap and his son Peter may be the child in the front row wearing a beaded hat with his hands in front of his chest. Indian Henry appears with his cloth hat in the back row, third position from the left. These Hän people had a camp downriver at Tthee t'äwdlenn/Eagle Bluff (near present-day Eagle Village) but spent several months in their brush shelters as nearby neighbors of the American surveyors. The Bancroft Library, Davidson Collection, Photograph Albums 1946.006 Alb, #17.

Dr. Kingsbury photographed his colleagues in front of the brush shelters of "David's Indians" in the fall of 1890 and entitled the image "Pirates of the Upper Yukon." The Bancroft Library, Davidson Collection, Photograph Albums 1946.006 Alb, #1.

wide-ranging travels of Chief Charley as well as of many other Indian men, lending credence to the idea that Big Paul could have been down at Charley's camp on the Kandik River in earlier years when the map was drawn, then moved upriver as the pace of prospecting and mining quickened around the Stewart and then the Forty Mile. While plausible, nothing in the records definitively casts Big Paul as Paul Kandik, and so the mystery continues.

What is clear is that this was a brief period of many meetings among very different groups of residents and visitors that generated some

Tom O'Brien, Jack McQuesten, and Al Mayo photographed by Dr. Kingsbury at their Forty Mile trading post in the spring of 1891. The Bancroft Library, Davidson Collection, Photograph Albums 1946.006 Alb, #26.

Children of Neehunilthonoh (Margaret Mayo) and trader Al Mayo photographed by Dr. Kingsbury at the Forty Mile trading post in the spring of 1891. The Bancroft Library, Davidson Collection, Photograph Albums 1946.006 Alb, #20.

intriguing photographs, stories, and documentary sources. More information about the surveyors' Hän neighbors may be available in other journals and correspondence of McGrath's party located at the National Archives in Washington, D.C. or elsewhere. The official Camp Davidson records are mixed in with thousands of pages of other correspondence and reports that form the U.S. Coast and Geodetic Survey records of Superintendent Mendenhall. These materials have not been microfilmed and are currently not available for research in the North. What remains is some remarkably clear visual evidence of these meetings and some inkling of their meanings, but only meager clues as to the identities of the Native people who lived out their lives in the area long after the surveyors departed for the south. So Paul the mapmaker might have been living near Forty Mile in the late 1880s and early 1890s, and may be in these photos, but no one alive today has made that identification and connection.

Government Records and the Gold Rush Era

Following Petroff's initial census of 1880, both the U.S. and Canadian governments conducted detailed population counts, with lists of names recorded for specific northern communities in Alaska and Yukon. The United States took censuses in 1890 and 1900, surveying individuals along the Yukon River at several points, including Circle, Charley River, Charley's Camp, and Eagle, listing many people with Native names and others with English names. Several people named Paul appear in these lists, with some even identified as pilots, but no one is listed as Paul Kandik, nor anyone with Kandik as a last name.[224] Some Yukon River people continued to travel north to the Porcupine River to trade at Rampart House after the H.B.C. moved there in 1870, but Paul Kandik does not appear in Canadian census records for the Porcupine River in 1891.[225] Numerous men named Paul appear in the 1901 Canadian Census but not Paul Kandik.[226]

Two more tantalizing glimpses of Native men named Paul occur in reports and records from the Klondike Gold Rush era and the years following. In the winter of 1897–1898 severe shortages of food supplies caused fear and concern among the thousands of people who had stampeded to the Klondike. The trading companies who supplied the region, along with both Canadian and U.S. government officials, feared that many of the newcomers would starve and/or become a serious threat to the security of the company warehouses and supplies for the whole community. To address these concerns a U.S. Army officer, Captain Ray, organized a winter evacuation of some stampeders who were without provisions and security for the various company warehouses. He had to make a very dangerous river trip in late fall from Eagle down to Fort Yukon and employed a man he called "Indian Paul" as guide and boatman.[227] Ray's party became trapped in ice when the temperature dropped suddenly and Paul was sent by land down to Fort Yukon to take a message to Ray's assistant there: "As soon as I landed I started the Indian Paul with a note to Lieutenant Richardson."[228] A nonnative man started out with Paul but soon returned, exhausted from traveling in the deep snow. Paul arrived at Fort Yukon four days later. Ray noted that an "Indian," unnamed in this section but perhaps again Paul, then brought supplies upriver to some miners who were stranded, saving the army officers "much trouble and labor."[229] This reference in Ray's 1897 report provides just a brief glimpse of another Paul from the Eagle area—likely a Hän man, skilled and willing to work with the newcomers who needed his help and knowledge. Is this Paul Kandik or Big Paul or perhaps Little Paul? Perhaps there are further details in some of Ray's journals or correspondence in the military archives in the United States.

One final reference to a Paul occurs in the journals of Anglican missionary Archdeacon Canham, who lived at Tanana in the late 1880s, then moved to establish the first mission at Fort Selkirk in 1894. Arthur Harper had established a trading post there, serving both Native customers and the increasing numbers of prospectors in the area. Canham, like Robert McDonald, kept a detailed daily journal. In 1903 he noted that an "Indian" named Paul drew a map for him of rivers, lakes, and trails in the area, including the Tlingit trade route between the coast and Fort Selkirk.[230] Is this the same Paul that McDonald baptized in 1887 at Harper's post on the Stewart? Did he later meet Canham at Tanana, and then relocate to Fort Selkirk after both Harper and Canham settled there in the 1890s? Canham provided no further details to identify this mapmaker named Paul at Fort Selkirk, so he remains an enigma as well. Several men named Paul are listed in the Canadian Census records for Fort Selkirk in 1901, but not named as Paul Kandik.[231]

Conclusion

For the time being none of the documentary sources beyond the Kandik Map identifies a person called Paul Kandik. The name Paul appears in several sources that could refer to the same man who made the map, but none of the chroniclers provides enough detail to be certain. Many Native people at the time clearly had the knowledge required to draw such a map and several did so, as noted in the records of nonnative travelers. These records all point to extensive and intensive sharing of geographical knowledge by Native people with both long-term resident traders and others like Petroff, Schwatka, and Allen who passed quickly through the area. The experience of Captain Ray with "Indian Paul" demonstrates the skills and willingness of Native people to help newcomers, a story told repeatedly before and after the Gold Rush.[232] While the documentary and oral sources for Paul Kandik are limited and tenuous, there is an extensive record for François Mercier, the other principal "author" of the map, providing a different perspective on pre–Gold Rush people, places, and events.

4

François Mercier: Agent of Change

In contrast to the search for Paul Kandik, the sources for François Mercier are numerous, including his own manuscript written after his departure from the North, now preserved at Gonzaga University Archives in Washington, D.C., and published eighty years after his death.[233] The Québec National Library and Archives holds numerous records on the Mercier family, photographs of François and his brother Moïse, plus published sources in English and French.[234] François is mentioned in Roman Catholic mission records located at archives in Canada and the United States,[235] in Anglican missionary reports,[236] in many accounts by his northern contemporaries, as well as articles in French and English newspapers of the time.[237] He was credited with providing the lettering of the names on the Kandik Map and with supplying geographical information about Alaska on maps published in France, while other sources acknowledged him as a leading northern expert.[238] Visual records include a romantic engraving of him in classic "coureur de bois" clothing with gun and snowshoes,[239] a photograph taken with Native associates at his Tanana trading post in the mid-1880s,[240] and another taken with nonnative colleagues as he departed the Yukon River for the last time in 1885.[241] Renowned French artist Alfred Boisseau painted an oil portrait of him in 1887, posed in fur parka and snowshoes, accompanied by an unidentified Indian man, with sled and dogs on the river ice against a hill resembling Tthee t'äwdlenn (Eagle Bluff) with a beautiful northern winter sky. Mercier is listed in Roman Catholic Alaska mission records as the father, or perhaps godfather, of William Taden (Dimoska), a child born to a woman named Mary Meldizun at Nulato in 1875.[242] So Mercier may have descendents somewhere along the Yukon today, although not identified with his last name or connected to his Québec family.

Although there are many references to his career as a trader and promoter, none provides a complete chronology of his activities or the many locations he occupied or visited during his seventeen years in the North.[243] There are significant gaps for some years and questions about where he was living, for how long, and his affiliations with various northern companies. There are also differences between the accounts of him recorded in English publications and those in French newspapers and books.[244] His own memoirs are sketchy on many details of his northern career. Most of the Alaska Commercial Company records, which might shed light on his activities, were destroyed in the great earthquake and fire at San Francisco in 1906.[245] As with Paul Kandik's story, Mercier's life and times in the North must be pieced together from the accounts of other traders, missionaries, and travelers, together with his own writing, and still there is much room for speculation and surmise about his motivations and activities between 1868 and 1885.

Mercier did not stay long in the North, but he initiated and participated in changes that had profound and lasting effects on the region. Despite this, he is not remembered in Native oral traditions, at least not among Hän people, nor does his name continue in usage as a personal, family, or place-name among residents of Alaska-Yukon today.[246] The Roman Catholic Mission at Eagle was named Saint Francis Xavier Mission, possibly as a reference to him.[247] Both François and his brother Moïse were celebrated in Québec newspapers in the late 1800s, and in later years their northern exploits were recalled by descendents of Moïse and publicized again in Québec.[248] François Mercier's great-nephews and nieces know some details of his fur-trading career, but little of the Native people he met and nothing of Paul Kandik. One great-nephew remembers a story about François from his father, who visited him in Montreal around 1900, but no one in the family has ever seen the North, where Mercier lived for more than a decade.[249] Some family members have recently seen a copy of the map and the published English translation of his *Recollections*, but they had never seen the original French version at Gonzaga University Archives, or the original map at the Bancroft Library.[250] Several researchers in Québec are familiar with the Mercier brothers' history, especially that of Moïse, who was for many years the mayor of Saint Véronique de Turgeon.[251] The present-day Yukon francophone community is keenly interested in his legacy and recently published a booklet with information about the Mercier brothers gathered from French newspapers and family sources.[252] Searching for François Mercier today is a far-reaching endeavor, with oral and documentary sources in French and English scattered across northern and southern Canada, the United States, and France.

François Mercier: Early Years in the American West and Alaska

François Mercier's origins and early career as a fur trader were described in an interview published in a French Canadian newspaper in 1871, while Mercier was visiting his family in Montreal. Written by an enthusiastic journalist, L. O. David, the article was a feature in *Galerie Nationale* and celebrated Mercier's fearless courage in traveling to far-flung regions of the American West as a very young man.[253] David saluted Mercier as a great man for adding to "*la gloire d'une nation*" and respect for "*le nom canadien*." The article named Mercier's birthplace as Saint Paul L'Ermite, a small town near Montreal where his father was a doctor. Leaving home at the age of eighteen he headed west, his imagination fueled by stories of French Canadian voyageurs in days gone by. He was at Saint Paul, Minnesota, for a few months, then joined the North West Company, trading furs between the border of Missouri and the Rockies. Living alternately at Fort Benton and Fort Union, he worked as a carpenter, gaining the confidence of his comrades and bosses as a capable and vigorous worker.[254]

These were dangerous times in the American West, with violent confrontations between Native people and advancing white explorers, traders, and settlers. The article referred to scalpings and massacres as a regular risk facing fur traders, including Mercier. He recounted his experiences at the siege of Fort Union, where he said that "six or seven hundred" Sioux attacked on horseback. According to the newspaper article, Mercier and his seven companions thwarted the attack by firing eight rounds and hitting eight Sioux in succession, after which their attackers fled. Despite this bloody episode the author insisted that French Canadians were well remembered by Native people throughout the continent, gaining admiration through "*leur douceur et leur bravure*"—gentleness and bravery. Mercier experienced many dramatic times during his years on the Plains, burying a young French Canadian companion who died of consumption, surviving a knife attack by a Native man, and trading stories as well as furs with ancient Elders.[255]

Mercier returned to Montreal in 1866 for two years, but the city life did not hold his interest and "*le gout des voyages*"—the taste for traveling—never left him. Forgetting the miseries and dangers he had experienced he set out again for new adventures in distant places, in keeping with his "national character" as a French Canadian. In California he dabbled briefly in mining. He woke up one morning, bored with this monotonous and sedentary life, to find that the United States had purchased Russian America, opening up a whole new "*vaste théâtre*" for the exploitation of the fur trade.

On April 15, 1868, filled "with joy," he departed from San Francisco on the sailing ship *Francis L. Steel*, bound for the Bering Strait and new adventures in the Far North. His companions were his brother Moïse, Emphrem Gravel of Saint Martin, Michel Laberge of Chateauguay, Napoleon Robert of Saint Césaire, and two unnamed Americans.[256]

After a rough two months at sea the party arrived on June 21 at the mouth of the Yukon. They built a small boat called "*la Canadienne*" and started up the river. Some four hundred miles upriver they were told that they were courting danger to go farther as no one had ever traveled up the next stretch of river. Undaunted, they continued, after telling the people that French Canadians never quit before attaining their goal. Mercier recounted stories of meeting many Native people in canoes, attracted by news of the voyageurs, some of whom he claimed had never seen a white person before. At the confluence of three rivers they built their winter quarters, called Fort Adams, on the Yukon River near the mouth of the Tanana. Native people from many points arrived to trade with the newcomers, and in spring they descended the Yukon with their boat loaded with furs. At the coast the owners of a rich new American company were established, against whom it would have been "impossible" to compete. The French Canadians sold all their goods to the Hutchinson, Kohl, & Co. representatives. According to I. O. David's article the new company offered so many advantages to François Mercier that he "consented" to join the firm as a trader.[257]

Returning upriver, Mercier encountered rumors of a possible attack by "Eskimos" around Nulato, and his group of Athabaskan guides refused to go on. Mercier seized two six-shot revolvers and stood in front of the group, asking whether they would rather face death at his hands or from the "Eskimos." His guides "naturally" preferred the latter and the party resumed their travels without further incident. Mercier must have had some doubts about the situation himself because he made out his will and left it at Nulato, telling the trader there to send it to his family if harm came his way. He spent three years on the Yukon with temperatures dropping at times to sixty below zero, then returned to Montreal sometime in 1871.[258]

As he prepared to leave his family in the spring of 1872 to travel back to Fort Adams, he told the reporter he would return to Québec in the fall. He intended to travel back and forth to the North for the next three years. The article concluded that passersby on the "Rue Notre-Dame" in Montreal would never guess that this civilized and impressive man was an "intrepid" northern fur trader, who had passed half of his life already among Native people, slept in the huts of "Eskimos," even eating at their tables—and these were not even a quarter of the tales that could be told by Monsieur Mercier! These were prophetic words indeed, as his life and times on the Yukon would be turbulent and filled with changes at every turn of the river.[259]

Time of Change: The Legacy of the Old Monopolies

The Alaska-Yukon trading scene that the Merciers joined in 1868 was a dynamic environment with distinct spheres of influence revolving around the former trading posts of the Russian-American Company (R.A.C.) and the Hudson's Bay Company (H.B.C.) The Russians had struggled against harsh climate and great distances to establish posts from Saint Michael's

Redoubt on the coast at Norton Sound to Sitka in the south, with a small adjacent hinterland extending as far as Nulato on the Yukon River and to outposts on the Kuskokwim and Copper rivers.[260] Russian traders married Native women and their children, referred to as Creoles, were employed as guides, interpreters, laborers, traders, and explorers, able to speak both Russian and Native languages.[261] The H.B.C. employed a similar strategy for establishing new posts and extending trade territory, with Métis men engaged as voyageurs in every party going west and north, who married local Native women and established families at frontier posts.[262] Family connections ensured access to geographic and community knowledge, together with translating and interpreting skills vital to business. H.B.C. traders arrived on the Yukon just twenty years prior to the Alaska Purchase, crossing over the mountains from the Mackenzie watershed in the 1840s. Alexander Hunter Murray established Fort Yukon at the mouth of the Porcupine in 1847 and Robert Campbell built Fort Selkirk at the mouth of the Pelly the next year. The H.B.C. also operated trading boats along the southeastern Alaskan coast after 1839, under terms of an agreement with the R.A.C.[263]

In between the two firms, Native traders acted as middlemen, hunters, and trappers, collecting furs, hides, meat, fish, and northern clothing from more remote neighbors and trading for European goods. As the R.A.C. and H.B.C. traders pushed farther into the hinterland that divided them, some Native middlemen lost their power and profits while others gained new influence, resulting in turbulent relationships, threats of violence, and some fatal incidents. Native and nonnative traders had some things in common—intense rivalries, fierce competition, fear of violence, and occasional bloody clashes, along with the hardships of a rugged environment, isolation, deprivation, and sometimes starvation. They were mutually dependent on one another for some goods and services, developed preferences, loyalties, and alliances, and engaged in ongoing exchanges of information about many aspects of the North—especially its geography, including the location and names for places and things in a variety of languages.[264]

The border between Russian and British territory was a matter of considerable interest to the company traders, but no barrier at all to Athabaskan traders. Murray carried some rudimentary survey instruments and knew when he crossed the 141st meridian on the Porcupine, and that he was far past the line at Fort Yukon. He worried about Russian hostility and located a site on the Porcupine in case he was compelled to "retreat upon our own territory." He heard about the Russians from upper Yukon people who traded downriver and also heard that some Native leaders objected to his presence in their country. Murray collected information to relay to company headquarters in eastern Canada and England. With the help of his interpreters, Murray constantly questioned his Native customers, finding out where they lived and traveled, whom they met, and what they traded.[265] The new fort, with its tobacco, calico, and other goods, was a magnet for Native people from near and far, disrupting previous trade networks, with consequent bad feelings and feuds. On the question of geography Murray had wide-ranging discussions, learning about the course of the Yukon from people who went downriver to the Russians and others who visited Robert Campbell's new posts upriver at Frances Lake and Fort Selkirk. He wrote to his superiors that "by questioning so many and comparing . . . statements, I have . . . some idea of the course of the Youcon and other rivers, of which hitherto very little was known. . . . I have seen very little of the Youcon, only a few miles above and below where we are.[266]

Robert Campbell traveled downriver from Fort Selkirk to visit Murray in 1850 and together they consolidated details on the course of the Yukon River. Campbell's post, located in the midst of Tlingit trading territory, was destroyed by Tlingit in 1852. H.B.C. officials focused on Fort Yukon, which was more profitable and easier to supply, and for the most part, acceptable to Native traders and therefore secure, despite its remote location. Maintaining that security and acceptance was the task of Murray's successors, Strachan Jones and James McDougall, who continued the practice of learning the languages, customs, and preferences of their customers.[267] Protestant and Roman Catholic missionaries arrived in the early 1860s, supported by the H.B.C. as a means of "civilizing" and educating Native people, reducing intergroup feuding and threats to white traders which interfered with the hunting and trapping required to maximize company profits. Protestant Factor James McDougall favored Robert McDonald of the Church Missionary Society over his Roman Catholic rival Père Séguin, supplying housing and food to the former in the cold winter of 1862–1863 while the latter was left to find shelter in Gwich'in lodges and nearly starved.[268] Over the next few years, McDonald's popularity expanded among Native people through his charismatic preaching and speedy acquisition of the Native languages. He traveled extensively with the people, going to their fish camps and hunting grounds, living in their skin lodges, sharing feasts of joy and occasions of sorrow. When scarlet fever devastated their camps he brought medicine, cut wood, hunted, cooked food, buried the dead, and comforted survivors, achieving a legendary status that had profound and lasting effects.[269]

McDonald was perfectly positioned to witness the next major change in the region, diligently chronicling the shifting trade scene and human relationships as the first American visitors arrived in the 1860s. He met the Western Union Exploring Party in 1866, noting the mix of nationalities that was a harbinger of things to come on the upper Yukon. American scientist Dall, British artist Whymper, Canadian explorers Ketchum and Laberge, and their Creole guide Lukeen, plus a number of Native paddlers from downriver groups, exchanged geographical information with the Fort Yukon residents that summer. McDonald learned about the downriver course of the Yukon from them, and he assisted the Canadians to prepare for an exploratory trip to Fort Selkirk.[270] Although Native paddlers had traveled the river for countless years and Lukeen had gone up to Fort Yukon in 1862, Dall claimed that his group was the first to explore the "terra incognita" between Russian and British territory, declaring that he was the first American to reach Fort Yukon from the sea: "This was the river I had read and dreamed of, which had seemed shrouded in mystery. . . . On its banks live thousands who . . . call themselves proudly, *men of the Yukon*."[271]

Dall's reports were among the first American publications on Alaska, circulated widely and cited for decades as authoritative sources on cultural and linguistic groups. He got help from Red Leggings, the chief at Fort Yukon, in collecting ethnological specimens and vocabularies for the various Native languages,[272] and from French Canadian Métis Antoine Houle, a "great favorite" whose house was always open to the Indians: "With them he could talk in their own dialects, while the usual mode of communication between the whites and Indians in this locality is a jargon somewhat like Chinook, known by the name of 'broken Slave.' The basis of this jargon, which includes many modified French and English words is the dialect of Liard River."[273] Dall enlisted an eclectic crew, which gives an idea of how many Native people were used to working with the Russians and ready for hire by anyone paying wages. At Fort Yukon Dall met Chief Shahnyaati', who was fascinating and at the same time repellent, with his many wives and reputation for evil threats and murder.[274] Dall reported that other tribes feared the "Tenana Kutchin," including the Hän or "wood people," and the "Gens de Bois" above Fort Yukon.[275] On the whole Dall's accounts were positive, stating that the Native people he met were friendly and very curious, willing to share information and eager to learn from the newcomers as well.[276]

The American Takeover: With French Canadian Traders

Rumors of change were flying up and down the river after the Alaska Purchase in 1867, with consequent uncertainties for many residents. McDonald wrote in October 1867 that the construction of a church at Fort Yukon had been "delayed owing to the uncertainty of the Hud. [*sic*] Bay Company's Fort on Youcon remaining where it is, as it is on what was till recently Russian American, but now United States Territory."[277] In the fall of 1868, he reported that three American fur companies had established posts on the Yukon and he worried that they were introducing "spiritous" liquors to the Native people.[278] As the Mercier brothers and friends traveled north from San Francisco in 1868 they were probably not expecting so much fierce competition so early in the American takeover in Alaska. Their Pioneer Company was just one of six small companies competing for furs on the coast and in the interior.[279] They all soon learned what their H.B.C. and Russian predecessors knew well. This was difficult country for travel and trade. They had to spend considerable energy and time hunting, fishing, gathering firewood, and doing other chores in order to survive, as did their Native customers, leaving only modest opportunities for trapping and trading furs.

The next few years were characterized by aggressive competition and constant change as the men and the companies they formed or worked for moved about, amalgamated, or quit the country. The Pioneer Company lasted for the 1868–1869 season only, with the Mercier brothers operating the Fort Adams Station fifteen miles down from the Tanana River, and other stations operated by five French Canadians, three Americans, one Englishman, and one Norwegian. The next year the Pioneer Company dissolved and the Mercier brothers joined two new companies. Moïse Mercier was hired by the Parrott Company to run Fort Yukon after the H.B.C. left. François Mercier joined Hutchinson, Kohl and Company (H.K.& Co.) and built a new post at the Tanana, since Fort Adams was in charge of the Parrott Company traders. His new post was called Tanana Station, located about twelve miles upstream from Fort Adams. The General Agent for H. K. & Co. at Saint Michael was a Finn named Captain Riedelle. The two Mercier brothers thus worked for rival companies briefly, though there is no indication of animosity between them in François' memoirs.[280]

In the summer of 1869 Robert McDonald was traveling as usual among the camps of Gwich'in along the Porcupine. He heard that American fur traders came to Fort Yukon "in a small steam-tug and with them two US Gov officers" who ordered the H.B.C. to leave after confirming this was American territory. Antoine Houle died in the fall—symbolizing the end of an era. McDonald planned to stay on the Yukon as the H.B.C. officers "made over the Fort" to him, but he knew he would have "to make some arrangement with US Commissioners to continue work." Moïse Mercier hosted him that winter at Fort Yukon, and on New Year's Eve he reflected on the situation: "The change [caused] much excitement among the Indians, and will prove a loss to them. . . . [It] hindered their autumn fur hunts. . . . [They plan] to abide with the H.B.Co. An offer is made to me to remain here as a guest of the American fur company . . . I do not think it will be desirable for me to do so."[281]

The next summer, in 1870, François Mercier observed McDonald's preaching when he arrived at the mouth of the Tanana where large groups of people from up and down the river had gathered as usual to trade. The missionary was accompanied by two chiefs whom he was training as Christian leaders—William Loola and Charles Titsishoori—both men of influence over their own bands and related groups. McDonald's charisma and popularity among these people must have impressed François Mercier because he offered to establish a mission at his Tanana trading post. Mercier suggested McDonald use his interpreter and invited McDonald to travel free of charge to Saint Michael and back in his company boat in early June.[282]

Great discussions were underway when they arrived at Saint Michael as the traders realized there was not enough business for two competing companies on the Yukon. McDonald reported that Captain Niebaum of H.K. & Co. and Captain Ennis of the Parrott Company "speak of uniting in to one company to mutual advantage" and operating only one post at Nuklukayet. François Mercier was present for these discussions. Later McDonald received "kind assistance" from Captain Riedelle, who assured him that no liquor was being imported and told him the companies were amalgamating in the best interests of all the traders to form the Alaska Commercial Company (A.C.C.). McDonald hoped the change would "benefit the Indians" as well. François Mercier visited McDonald as he was carrying on his missionary work at Saint Michael, attending prayer services with the Indians.[283]

On July 21 Mercier and the other traders departed for the upper Yukon on their small steamboat for the six-week voyage to Fort Yukon. McDonald was on board too and was a keen observer of the traders, noting where each one was left for the winter with trade goods in the new network established by the amalgamated company. No doubt the traders were paying close attention to McDonald as well since he was attracting Native people everywhere he went. Along the way he worked on some "Eskimo translations" with Mercier's Creole interpreter, Timouski. At every stop people gathered to meet him as word of the preacher spread up and down the river in a few short weeks. At a camp above Anvik the people "knew a hymn that I taught to the Indians at Nulato; they were taught it by some there."[284] He met some Nuklukayet people who had been down to the coast to trade and were paddling all the way back home in their canoes. From others he heard news from above Fort Yukon that "Nootleti," chief of the Hän, had died. The moccasin telegraph was very effective as Native people were constantly on the move trading and visiting at the various posts and their own camps.[285]

Just above Nulato the traders met Moïse Mercier and his partner M. Dufresne, who were hurrying downriver with the alarming news that "the hunkutchin [were] threatening to expel the Americans from Fort Youcon."[286] As the little steamboat advanced upriver, the Merciers and their partners must have pondered what course to take at a frontier post, thousands of miles from any military assistance, among people who were clearly unhappy with the changes they had introduced. At Nuklukayet, Captain Riedelle announced to McDonald that the steamboat could not go farther upstream because of engine "problems," but the situation at Fort Yukon probably influenced that decision as well. Riedelle collected "all the goods here and as much of dried and salted salmon as could be taken in the boat" and departed for Fort Adams, where François Mercier was in charge. There was to be no company post at Fort Yukon, so it is likely that the two Merciers were together that winter of 1870–1871.[287]

McDonald continued up to Fort Yukon where James McDougall was waiting for him, apparently not in any danger from the Gwich'in and Hän who still preferred their H.B.C. connections and included the missionary in the favored circle. Together they traveled up to Rampart House for the winter. McDonald must have considered it useful to maintain friendly relationships with the American traders, since he sent his map of British North America to Captain Riedelle via his paddlers who returned to Nuklukayet.[288] The Merciers had much to contemplate during the long winter at Tanana. Local Gwich'in or Hän people plundered their property at Fort Yukon again. McDonald wrote: "The poor Indians! Very little can be said to palliate their offence. They were ill-disposed to the Americans, since they regarded them as the cause of Fort Youcon being abandoned by the H.B.C., in which they felt they suffered a loss."[289] In July 1871 McDonald saw evidence of the destruction in person, stopping at the fort on his way to the upper Yukon to visit the Hän. The place "looked desolate" with the buildings in ruins. He left a note in one of the buildings for Captain Riedelle, hoping the A.C.C. boat might arrive sometime soon. It was unanswered when he returned so the American company traders probably did not venture back to the post that summer.[290] According to Mercier Mrs. Riedelle had arrived at Saint Michael, so the Captain may have canceled his annual visit in view of the hostile situation the previous year. Mercier left for Montreal, perhaps discouraged by these events and considering whether to return to a place where the H.B.C. still held sway.[291]

Meanwhile Robert McDonald had great success with his missionary work in the region. Traveling with William Loola and Thomas Koisa, he was welcomed warmly at Hän camps on the upper Yukon. At one camp a mix of Hän and Kutchakutchin families practiced prayers and hymns with Lucy, wife of Thomas Bear. At other camps people listened well but McDonald feared they "did not understand much as I had no interpreter . . . Charles Titsiyoorzi interpreted for me at the first camp of hun."[292] Further upriver he visited a camp of the "Trohstik-kutchin, one of whom is Katza, formerly an interpreter of Mr. Campbell, HBC . . . [who] kindly interprets for me."[293] McDonald completed work on a writing system and translations of religious texts into the Tukudh language, and in the spring of 1872 he left for the south, traveling to England to supervise the publication of the first Tukudh syllabarium.[294] His brother Kenneth spent the winter of 1872–1873 at Rampart House, traveling to the Gwich'in and Hän camps to teach Christian doctrine and lessons in English.[295] The McDonald brothers attracted people from the Porcupine and the Yukon to Rampart House, where they traded with their old H.B.C. friends at the same time. No doubt they conveyed news of their travels when they visited the Merciers at the Tanana, comparing prices and bartering for deals as they had done with H.B.C. and Russian traders in the past.

François Mercier had a new opportunity to try his hand at advancing the A.C.C. trade when he returned to Saint Michael in the summer of 1872. Captain Riedelle was ill and left for San Francisco. Mercier was appointed as the new General Agent of the Company for the Saint Michael or Yukon District, which he described as comprising "the Youkon, and the Kuskokwim . . . more than two-thirds of all the territory of Alaska . . . near 600,000 square miles . . . three times the size of France . . . [with] only 34 white men residing, counting the Russians, and all except two who resided on the Kuskokwim, were dispersed . . . along the Youkon."[296] Mercier listed all the traders at various posts with nationalistic pride: "As one can see by this list, French Canadians have been put in charge of all the principal trading stations of Saint Michel, while those of 2nd or 3rd rank were given to other nationalities."[297] Mercier decided to reoccupy Fort Yukon that summer, placing his brother Moïse in charge, while he most likely was at Saint Michael for the year.[298]

During that winter of 1872–1873 Moïse hosted some important visitors, the result of letters François had written to their family in Québec lamenting the fact that there were no Catholic missionaries in his region. Having failed to woo McDonald with offers of a Protestant mission at the Tanana, and perhaps suspicious of his continuing loyalties to the H.B.C. and influence among the Indians, François was pleased by the arrival of Monsignor Clut from the Mackenzie region in the fall of 1872.[299] The Fort Yukon Gwich'in, led by Chief Shahnyaati', ignored the priests out of loyalty to McDonald. French was the language heard most often that winter at the fort, since all the nonnative residents were francophone. In the spring the priests moved downriver to find more rewarding ground for their cause. François was enthusiastic about their arrival at the Tanana and took them downriver on his annual trip, "stopping at one village after another along the Yukon . . . to baptize a large number of children, of which I was almost always the godfather."[300] Overall the Alaska-Yukon adventure of Monsignor Clut was a success and Mercier was proud of his role. McDonald gleefully reported to his sponsors that "a Romish priest" had passed the winter at Fort Yukon with the American traders, but had "not succeeded in 'seducing' any of the Indians."[301]

More Change on the Yukon

In the summer of 1873 Mercier welcomed a second set of newcomers from the South who were also seeking to make their mark, though in a very different way. From the mid-1860s word of successful gold mines in the Cassiar District of British Columbia drew prospectors from California and other southern diggings north in search of new ground. Some continued further up the Alaskan coast to Sitka, while others traveled the old H.B.C. routes down the Mackenzie, over to the Porcupine and the Yukon. The first of these hardy individuals, Arthur Harper, Fred Hart, George Finch, and Andrew Kenseller, arrived at Fort Yukon in May 1873.[302] The first three went to prospect around the White River the next winter, while Kenseller stayed at the A.C.C. post with Moïse Mercier. When François arrived in August "to take stock of the needs of the trading post [he] hired [Kenseller] on the spot to be the assistant of Napoleon Robert at the Noukelakayet station."[303] After François left to return to Saint Michael, a second prospecting group arrived at Fort Yukon, consisting of Leroy (Jack) McQuesten, Al Mayo, and George Wilkinson.[304]

All of these prospectors had many connections to the North before they arrived.[305] According to McQuesten they "heard a great deal about the Yukon River" from H.B.C. men they met while they were trapping around Hay River in 1871 and decided "we would go and see for ourselves what the country was like." They went up the Liard and wintered on the Nelson River in 1872, meeting "James Sibistone [*sic*]" who had been at Fort Yukon when Captain Raymond visited in 1869. He told McQuesten about the wealth of furs on the Porcupine, Yukon, and Tanana rivers and also that "one of the officers that came up on the steamer washed out a jar of dirt near Fort Yukon and he had about a teaspooneful [*sic*] of something yellow in the pan and the officer threw it away remarking that it would not do to let the men see it as they would all leave the steamer." McQuesten wrote, "These stories interested us; it was just the country we were looking for and it gave us the encouragement to make the trip." James McDougall, the factor at nearby Fort Liard, visited McQuesten and Mayo, confirming the stories and offering "to set them up" in the Porcupine country with boat lumber, a guide, and provisions, no doubt hoping to add these experienced men to his orbit rather than see them become free competitors or allied with Mercier's operations on the Yukon.[306]

When McQuesten and Mayo arrived at Fort Yukon on August 15, 1873, they met Moïse Mercier. It was late in the season and supplies were low: "Mercier had no flour to sell—he let us have 50 lbs. That was for four of us for one year. We left to go to our winter quarters about one hundred miles below to some lakes that McDougall recommended to us. Near the mouth of Beaver River we passed the place and went into the Ramparts. Then we found some Indians—they made us a map and then we went back."[307] The men built a cabin, hunted moose, bear, rabbits, geese, and ducks in the fall and fished in the winter. They expected to trade for furs but never saw "one Indian all winter" and concluded furs were scarce. After Christmas Mayo and McQuesten traveled up to Fort Yukon. Arriving with "one fish left and faces frozen," they found Moïse Mercier "as happy as when we left him in the summer." They were "treated to the best there was" by the trader and decided to move to the post. McQuesten traveled up to Lapierre House where he saw "Sibbestone that told us such big stories about the Yukon. We made fun of his fox and gold stories but he swore it was the truth but we had not looked in the right place."[308] McQuesten may have been checking out his options for obtaining support as promised by McDougall the year before but came back down to Fort Yukon.

When the ice went out in June 1874 Moïse loaded up his boat and took all the prospectors down to Saint Michael where "Frank" Mercier, as McQuesten called him, was stationed as the A.C.C. agent. They were all treated "as if ACC employees," and after the company schooner arrived and supplies were unloaded, Mercier hired Hart, Mayo, and McQuesten while Harper, Finch, and Wilkinson decided to prospect again that year.[309] Mercier required McQuesten to sell him "the furs that he had bought from the Indians (about 100 marten and about 20 beaver) as well as the merchandise that he had left at Fort Yukon in the care of my brother which he had obtained from the Hudson Bay Company . . . which came to about

$1400."[310] McQuesten was recommended by Monsignor Clut "as being a very honest man and a good fur trader" so Mercier assigned him to a new post far up the Yukon.[311] It is interesting that Mercier selected a new man, and an English speaker, rather than one of his more experienced French Canadian traders or even his brother for this enterprise. Perhaps he calculated that a new trader who spoke the same language as McDonald and most of the H.B.C. men on the Porcupine would be more successful with the upper Yukon Hän. Mercier no doubt knew that he could drive a hard bargain with McQuesten after his first grueling winter. Finding himself in the midst of two monopolies, with the H.B.C. on the Porcupine and the A.C.C. on the Yukon, McQuesten could neither sell his furs at a decent price, nor ship them out, nor resupply himself for another winter, without the support of one of the two companies.

Fort Reliance was Mercier's most ambitious undertaking in the North as Agent of the A.C.C., pushing the company's network some three hundred miles above Fort Yukon into territory where the H.B.C. had not ventured since Campbell had passed by twenty years earlier. Mercier gave the following explanation for his decision: "It is that before 1874, a number of years before the discovery of the gold mines, when there was only a question of furs, the Indians of the upper Youkon, and the more numerous Indians of the upper Tanana, were obliged in order to exchange their furs to take them to Fort Yukon or the Noukelakayet station, situated several hundred miles from their hunting grounds (500 or 600 miles). It was, then, to shorten at least by half this long and arduous journey which the Indians had to make every spring [that Fort Reliance was built]."[312] The trip on the little steamer *Yukon* required a whole month of difficult travel, which would have been "less than half the time, if this little steamboat had not been hauling three small barges loaded with provisions and trading merchandise for Fort Nulato, Noukelakayet, and Fort Youkon. It was also necessary for us to stop at least six hours every day to cut wood to burn in the boiler of the little steamboat."[313]

While Mercier may have been thinking primarily of the Indians and their furs, his new employees and their associates had prospecting on their minds and his decision to push up the Yukon placed them in the heart of promising new ground. Ironically, discovering gold would prove as elusive for them as success in the fur trade was for François Mercier. In mid-July the A.C.C. steamer reached Tanana, where Harper and several others left to prospect, while Mercier continued upriver with the rest of the group, guided by Chief Catsah[314] and ten other "Trondiak" men who were on board the steamer.[315] They arrived at the site for the new post on August 8 and Mercier left on August 9 "after having cleared the place . . . by the crew of the steamboat."[316] McQuesten supplied more details in his *Recollections*: "We selected a location near Trundeck about 350 miles from Fort Yukon. . . . I employed some Indians to carry logs and some went out to hunt. . . . We had our house and the store completed and the Indians brought in plenty of dried meat to last us all winter. I sold all the goods we had for furs during the winter."[317]

Neither Mercier nor anyone else commented on the risks associated with this new post after the destruction at Fort Yukon just two years before, attributed to unspecified Hän people. Had François and or Moïse been able to placate the Hän through better trade offers in the intervening years? Had their absence from Fort Yukon caused a change of heart among Native traders who had to travel farther to get American goods at Tanana? Were the attacks the work of particular groups of Hän or Gwich'in, perhaps people from Charley's camp who were closer to the H.B.C. traders on the Porcupine and allied with Shahnyaati'? Catsah had been Robert Campbell's interpreter so he may have been keen to see a trading post right across from his own camp, where his people could easily acquire goods, hunt for the white men, and learn more about their ways. It was certainly a bold move on the part of Mercier, and perhaps Catsah as well. As for McQuesten and Banfield, they embarked on a grand adventure, the only white men living hundreds of miles from their compatriots, with no dogs, canoes, or other means of transport of their own, completely dependent on their new Hän neighbors until the steamer returned the following summer.

François returned to Saint Michael to bid adieu to Moïse, who left for his Québec home in late summer, never to return.[318] François spent the winter of 1874–1875 at Saint Michael, then took the company steamboat as usual upriver to Fort Yukon in the summer of 1875. McQuesten reported that "Mr. Mercier made a good trade in furs" and they all went to Tanana Station, where Harper and the other prospectors had found gold but nothing in paying quantities. The whole group went downriver to Saint Michael where Mercier was in for a surprise. McQuesten reported that "everything changed in the business. Mr. Rudolf Neuman was appointed

ACC Agent for the district and the 'Station' [i.e. Tanana] was to be let out to the traders on percentage. A. Mayo, Harper and myself had from Fort Yukon and the upper country. Harper and Mayo went to Fort Reliance and I stopped at Fort Yukon."[319] Mercier's roles and whereabouts for the next two years are a mystery. Did he voluntarily give up his position as A.C.C. district agent? Was he replaced owing to a failure to meet company expectations for profit margins or some other breach of policy? Did he quit the company and follow his brother home to Québec, or simply go as far as San Francisco to seek out new opportunities for sponsorship? He may have chosen, or been assigned, to go to a post on the Kuskokwim at this time since he later claimed to have traveled on that river sometime in his career.[320] With Moïse gone from Fort Yukon and François gone from the Tanana, Harper, Mayo, and McQuesten were left as the most experienced upper river traders and in charge of the whole region, only two years after their arrival.

Harper and Mayo had already married Native women and McQuesten had met the woman who would marry him in a few years so they were settling in for a long stay in the North. With their wives, they were well placed and connected to Native trappers, hunters, paddlers, and other help that was absolutely critical to success in their business. The three men formed some type of company at this time, in order to carry on business in the vast territory assigned to them.[321] With a prior friendship, common interest in prospecting, and English as their primary language, they may not have been interested in having Mercier join them as a partner, or perhaps he was not interested in working as an equal with his former employees. This was the year that Mercier was listed as the father or godfather of a child born at Nulato, so perhaps he was located downriver, trading there or nearby.[322]

François Mercier arrived back at the Tanana in the summer of 1877, building a new post he named Fort Mercier, trading for the "opposition" Western Fur and Trading Company (W.F.T.C.),[323] and paying very high prices to attract customers. McQuesten reported that Mercier's tactics meant there was "not much to be made on percentage" by his group of independent traders working on commission for the A.C.C.[324] With two rival American posts at the Tanana, posts at Fort Yukon and Fort Reliance, and the H.B.C. at Rampart House, the Native traders had plenty of options for bargaining and obtaining the best deals. The situation must have caused some tensions among the white traders, as they were all operating on a shoestring in remote and difficult country, but their narratives written years later provide sparse evidence of their feelings. Certainly Mercier and the A.C.C. traders were on friendly enough terms to visit each other at their posts and even travel together on occasion.

During the summer of 1878 trouble broke out on the Tanana and the rival traders worked together to take care of a serious problem. Mercier was at his post while Harper and Mayo were at Tanana Station a few miles downriver. Mercier's former Parrott Company associate James Bean arrived in June and went up the Tanana as an independent trader, taking his wife and small child with him. Mercier and others tried to dissuade Bean but he persisted. That fall some Tanana men killed his wife, an event that Mercier sadly recorded as another "first" in the county: "Mrs. Bean was

Charles Farciot photographed this Tanana woman after she returned from hunting c. 1883, probably at Nuklukayet. Alaska State Library, Wickersham Collection, Charles Farciot, SL-P277-017-011.

the first white woman who . . . went up the Yukon as far as Noukelakayet, as well as the Tanana River, to about 30 miles above the mouth. The first who brought into the world a pure white child in this country, the first who lost her life there. The little station . . . on the Tanana was . . . the first . . . occupied by whites on this large river."[325] McQuesten gave similar details commenting on the status of trade and exploration on the Tanana at that point: "We advised him [Bean] to leave his family as the Tanana Indians were not very friendly . . . although that winter [1877–1878] I was up the River 350 miles with three Indians . . . and traded a large amount of furs. . . . the first whiteman. . . in their country and it was to their interest to treat me well as they found it much more convenient to have goods brought to their camp than it was to go after them."[326]

This violent incident played on the traders' minds for years to come and must have created considerable discomfort among Native and non-native people at the Tanana Station. Harper and Mayo were both living there with their Koyukon wives, so presumably had apprehensions for their families, and perhaps also some protection if other relatives were living close by and employed by them at their post. Mercier did not mention any family or associates at his post by name, but he certainly would have had an interpreter and other Native guides, hunters, and paddlers nearby. Harper made the trip up the Tanana to get Mrs. Bean's body, not Mercier, though he had been in the area longer at that point and must have known the Tanana people as well as or better than the other traders.[327] Possibly Harper was accompanied by some of his wife's relatives and was more secure in undertaking the trip to the murder scene.

Mercier's trade practices continued to have a negative economic impact on the other traders at the Tanana. As a result, McQuesten returned to Fort Reliance in the fall of 1878, prospecting in the Sixtymile during the winter and finding small amounts of gold.[328] Mercier must have impressed the owners of the W.F.T.C. with the potential for trade on the upper Yukon and Tanana because his firm brought in a small steamboat in 1879 called the *St. Michael*.[329] Mercier did not write much about this period but McQuesten said the new boat came as far as Fort Yukon that fall and "they laid her up for the winter. Captain Anderson was in charge, Weaver was Engineer, Waldron was Factor in the fall of 1879."[330] Mercier's role in these operations is unclear, but it seems likely he concentrated his efforts on the Tanana fur trade, perhaps even exploring the river during this time.

He credited Bean with building the first post on the river in 1878, and McQuesten claimed to be the first white man to have traveled up the river in 1877–1878, so Mercier was not the first white trader on the river, but he may have made winter or summer excursions, before or after the arrival of the *St. Michael.* On Reclus' map Mercier's name is shown on the upper Tanana, most of which was not navigable by steamboat because of rapids, so if he went there he traveled on foot or by canoe.[331]

Mercier had some problems with the local Native people at his post, caused by a serious misunderstanding of their beliefs and customs. He did not relate this incident in his own memoirs so it is McQuesten's

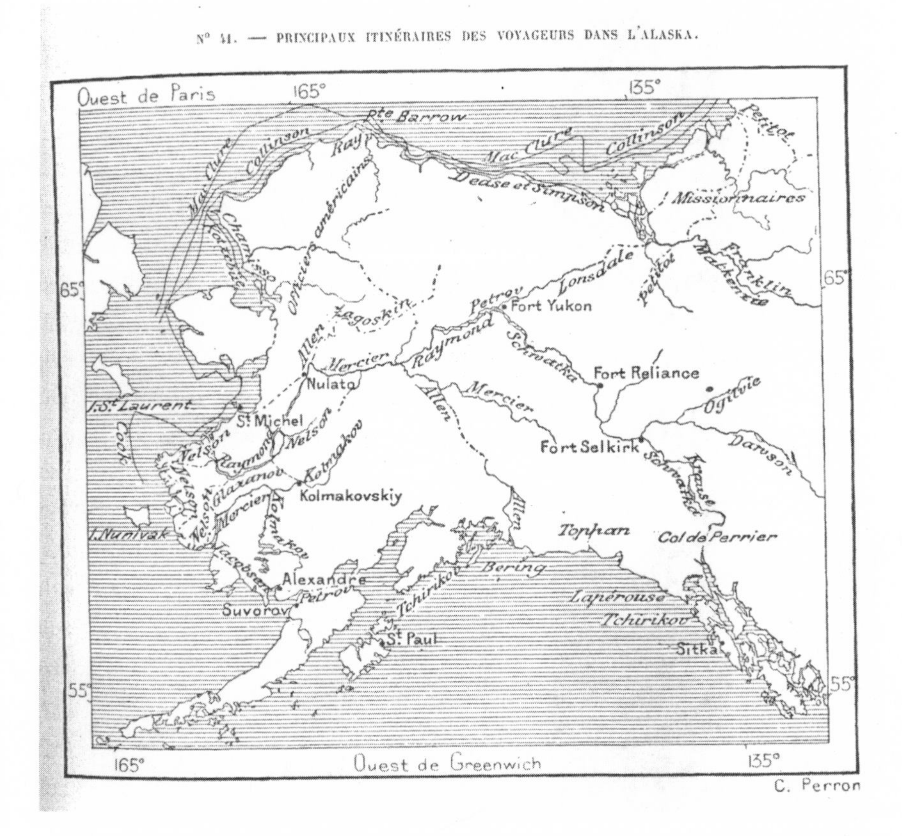

In 1890 French geographer Élisée Reclus published a geography book on North America with a map of explorers in Alaska and the Yukon. François Mercier's name appeared on the middle section of the Yukon River, on the upper Tanana River, and on the lower Kuskokwim River. Élisée Reclus, *Nouvelle Géographie Universelle La Terre et Les Hommes xv Amérique Boréale*, 195.

Recollections again that supply the details: "Mercier . . .bribed an old man to get him a Whiskey Jack's nest with their young birds. . . . The Indians are very superstitious about robbing these nests but a sack of flour and some tea and sugar and some calico to the old lady overcame the old man's scruples and he found a nest and gave it to Mercier. Shortly after it turned cold and remained so long that the Indians began to make medicine to break the ice."[332] Threatened with banishment by his band for breaking this taboo the old man sought protection from Mercier. The incident was finally settled peacefully after Mercier gave a feast for the local people to make amends.[333]

The Summer of 1880: The Census and the Kandik Map

In the summer of 1880, the year the Kandik Map was drawn, Mercier was still employed by the Western Fur and Trading Co., and as usual, spring found him at Saint Michael with his furs, just in time to meet Ivan Petroff. Mercier had very little to say about Petroff, though the two had French in common and a mutual acquaintance in Bishop Seghers.[334] He noted only that Petroff was "sent by the United States government to take a census of population of the Territory, visited the Youkon from its mouth at the sea to Noukelakayet, and also part of the Kuskokwim River."[335] Petroff's annotation on the map states that it was received from Mercier, but he never mentioned the trader by name in his correspondence or reports. He thanked both the A.C.C. and the W.F.T.C. for "free passage . . . to and from the Territory and other courtesies extended."[336] He offered a few pithy observations on activities at Saint Michael: "Since I arrived here I have been sleeping on a bearskin—beds are scarce—but we get very little sleep just now and it is all daylight. The traders who come down here once a year from their stations 1600 miles away do not care about sleeping much and business is going on night and day."[337] The only trader he mentioned by name in his letters to his wife was Jack McQuesten, perhaps because he had arranged to travel with him on the A.C.C. steamer upriver.[338]

Although he relied on traders for his information about the interior, Petroff expressed frustration at the lack of detail they could or would provide: "A few traders and prospectors have gone up the Tanana and over the old established track of the Yukon. . . . But the trader sees nothing, remembers nothing but his trade and rarely is he capable of giving any definite information beyond the single item of his losses or his gains through the regions he may traverse." He did not say which traders were on the Tanana but it seems likely that if McQuesten and Harper went up the river, Mercier would have pushed his "opposition" company's trade there as well to maintain his competitive position, though whether he or an assistant actually did the traveling is open to question, with no accounts to corroborate his presence on the river. Petroff noted that the A.C.C. was no longer the only firm trading in the region and rival traders dazzled the Native hunters "with lavish display of costly articles of luxury and delicacies for the palate, exciting them to the utmost exertion in the pursuit of fur bearing animals." The competition caused the firms to operate at a loss in some years.[339]

Through the traders, their interpreters, and the Native people he met, Petroff heard that much of the interior was still unexplored by nonnative travelers and the available information was unreliable: "What the country north of Cook Inlet is like no civilized man can tell, as in all the years of occupation of the coast by the Caucasian race it has remained a sealed book. The Indians tell us that the rivers lead into lakes . . . [and] flow into . . . the Tanana and the Yukon . . . with stories of mountains of immense altitude visible for hundreds of miles . . . which may be accepted until reliable explorers . . . penetrate this region. . . . The most important tributary . . . is the Tanana, the river of the mountains . . . its headwaters approaching the Upper Yukon within five or six days 'Indian' travel. . . . [S]amples of surface gold . . . have been exhibited. Of the upper . . . Kuskokwim River I have no authentic reports, but the natives relate that along its several branches the country is a level plain encircled on all sides by tremendous mountains."[340]

While Mercier apparently was the source of some of this information and provided the Kandik Map at the outset of his trip, Petroff must have received more help from the A.C.C. traders as they were all named in his 1882 report. In many ways Mercier might have seemed a more natural collaborator, with his sophisticated Québec background, than the prospectors with their Native wives and families. However Petroff was homesick on his long trip[341] and he may have warmed to the hospitality and family life of the Harpers and Mayos at Tanana Station. As for the Kandik Map and its apparent multiple sources of information, this explanation seems to fit well too. Petroff no doubt showed the map to the prospector-traders and their extended families, all of whom would have been well equipped to share more details with him than either François Mercier or Paul Kandik on their own could have provided, especially for the upper Tanana and Kuskokwim connections.

Back to the Hän Country Again

The work of a government agent was probably the least important of François Mercier's concerns that summer as he had ambitious plans for establishing yet another new post on the upper Yukon, in direct competition with McQuesten and the A.C.C. at Fort Reliance. He took the *St. Michael* upriver, only his second trip above Fort Yukon, and the first time he visited the Hän people since his brother had left six years earlier. He may have been traveling with new Hän guides this time, since Chief Catsah apparently died around 1880.[342] Mercier established his new enterprise "80 miles downstream from Fort Reliance on the left bank of the Youkon, about ¾ miles between 'David's Village' (an Indian village of the Gens de Fous) and the mouth of the little Tatotlin River," at the present site of Eagle.[343] He left Captain Anderson in charge for the winter of 1880–1881 and returned to Saint Michael. Indians had reported white men on the headwaters of the Yukon that summer. McQuesten "supposed it was prospectors looking for gold and it proved to be the case as Densmore and Slim Jim were in about that time."[344] Prospecting was still on Harper's mind, too, and the next summer of 1881 when the *St. Michael* arrived at the post, Mercier reported that "Mr. Beate [or Bates as McQuesten named him], shareholder in the W.F.&T. Co . . . in company of Mr. Harper . . . traversed the region that here separates the Youkon from the upper Tanana."[345] Harper and Bates followed one of the trails marked on the Kandik Map with some local Indian guides, very possibly encouraged by all the discussions occasioned by the drawing of the map for Petroff the year before.

Just as prospecting was beginning to look more promising, the fur trade was as precarious as always, with McQuesten reporting that the "opposition" traders on the *St. Michael* had traded very few furs that year on the trip upriver. After just one season at David's Village "Captain

Anderson was so disgusted with the Indians that he took his windows and stove out of his house and abandoned the station."[346] Mercier may have been trading still for the W.F.T.C. at his Fort Mercier post near the Tanana during 1881–1882, but made no mention of that season in his memoirs. In the summer of 1882 he came back to the site near David's Village and "reestablished it on his own behalf while trading on commission for the Alaska Commercial Company, naming it Belle Isle Station, the name of one of his friends from San Francisco."[347] Mercier built yet another new house there as another "opposition" trader took over Captain Anderson's cabin. McQuesten reported, "The Indians had a picnic that winter as they both had quite a supply of flour and they were both out in the spring."[348]

Many more prospectors entered the country after Captain Beardslee negotiated the opening of the Chilkoot Pass with the Tlingit chiefs in 1881. This shortened the voyage to the upper Yukon by many weeks, allowing prospectors to come in for the summer and return to the coast for the winter. In the winter of 1882–1883 some gold seekers stayed for the whole winter, encouraged by McQuesten at Fort Reliance, who had plenty of flour and goods and was ready for company. Joe Ladue, William Moore, and another group of seven all built cabins, to the delight of the trader: "It was the first time with the exception of one year, that anyone was living near that I could converse with."[349] Mercier was downriver at Belle Isle and surprised by the arrival of four visitors in April "and the foremost gave me his hand, addressing me in French . . . Joseph Ladue, French-Canadian [and three others] . . . were the first whites who entered Alaska by the route of the Chilkout Trail and Upper Youkon . . . up to that date, we were under the impression that Fort Reliance was found in Alaska, though it was really in British Columbia."[350]

Information about Mercier's final two years in Alaska is scarce, with no detail provided in his memoirs. The summer of 1883 was a year of more dramatic news as the W.F.T.C. sold out to the A.C.C. and many prospectors left, perhaps being wary of the A.C.C. monopoly on the Yukon. Belle Isle and Fort Reliance were abandoned as all the traders and prospectors went to Tanana to winter.[351] Mercier may have been the A.C.C. representative at Nuklukayet since the McQuesten group had become independent traders, buying their own steamboat, *New Racket*, from the Schieffelin Brothers, who had left after one winter of prospecting.

Charles Farciot photographed these three steamboats belonging to rival trading companies at Nuklukayet in the spring c. 1883. From left to right: the *St. Michael*, the *New Racket*, and the *Yukon*. Alaska State Library, Wickersham Collection, Charles Farciot, SL-P277-017-032.

In this photo Charles Farciot shows the steamers *New Racket* and *Yukon* pulled up on shore with people walking on the frozen Yukon River in front of the Anglican Saint James Mission at Nuklukayet, c. 1883. Alaska State Library, Wickersham Collection, Charles Farciot, SL-P277-017-027.

Willis Everette reported that François was upriver at Fort Mercier in the summer of 1884 when he stopped to visit him there.[352] Mercier's friend Charles Farciot photographed him there on his front porch about the same time, just before the building was dismantled and moved down to Nuklukayet.[353] Mercier must have had regrets on that occasion as he considered his namesake post to be, not surprisingly, "the best constructed on the Youkon."[354] An annotation on the copy of this photo in the Wickersham Collection reads "Fred Mercier's Family," while the caption on the same photo published in Québec in *Le Mond Illustré* in 1897 lists the names of all the adults and the dog, but not the children in the photo, leaving room for speculation as to whether any of them were Mercier's offspring.

François Mercier at his Fort Mercier trading post at the Tanana c. 1884. A handwritten note on the photo in the Wickersham Album reads "Fred Mercier's Family." Mercier likely provided the identifications of the people published in a Québec newspaper article in 1897 that included the same photo. The information in the article is translated into English and indicated in quotation marks. From left to right (seated at left end of porch): unidentified man, "Chief Starcke Souhague," unidentified man, (standing on porch) "Androuska, Russian Métis," his wife "Matrona, American Métis," "Mr. François Mercier with his faithful dog Jack," unidentified girl, unidentified girl, "Sport, Chief of the Nowakaket Indians," Mercier's nephew "J. Baudoin" with unidentified girl, unidentified man, "Siroschka," unidentified child, unidentified child. Alaska State Library, Wickersham Collection, Charles Farciot, SL-P277-017-037. Identification of people from Picard, "Souvenirs du Voyages," *Le Monde Illustré,* September 4, 1897, 294.

The winter of 1884–1885 was cold, with temperatures plummeting to eighty below zero.[355] The cold may have factored in Mercier's decision to leave the North the next summer. He wrote only that Dr. Everette "returned to civilization on the steamer St. Paul. I was on the same boat."[356] Of the fur trade scene he simply commented that "all the employed fur traders who were still living in the district of St. Michel in 1885 when I left the country, as well as all the companies . . . were amalgamated into the Alaska Commercial Company, the foremost trading company in the land."[357]

Conclusion

François Mercier was in his seventeenth year of trading in the North when he departed. With financial resources and established family in Québec, he was the only trader who could and did travel south for holidays. He spent at least one and possibly more winters outside, but he could still claim many years on the Yukon and at Saint Michael, living most of his adult years there from the age of thirty-four to forty-eight. He was an independent trader at various times, a trader and key promoter for the Western

Charles Farciot took this photograph of a group traveling by dog team in the vicinity of Nuklukayet when the temperatures dropped to fifty below zero c. 1883. Alaska State Library, Wickersham Collection, Charles Farciot, SL-P277-017-043.

Fur and Trading Company, a trader and the district agent for the Alaska Commercial Company, and probably finished his northern career as the company's trader at Nuklukayet Station, formerly called Fort Adams, exactly where he started out in 1868 with his brother and the other Pioneer Company men. He was constantly pushing forward with new ideas and ambitious building plans, but his success was curtailed by the whims of corporate change, through decisions made in San Francisco by company shareholders, and perhaps by his own approaches and attitudes. Success always seemed to be just beyond his grasp, with others often reaping the rewards and accolades that flowed from the foundations he laid.

Initially his French Canadian background and western American frontier experience may have played well in the North, as a transition from the old H.B.C. regime with its French-speaking Métis voyageurs. Their connections to Native families introduced a francophone element at Fort Yukon and surrounding areas. In the early American period, francophone influence faded when the H.B.C. withdrew and Moïse Mercier departed.

This winter photograph of the trading post at Nuklukayet c. 1883 shows the steamers *New Racket* and *Yukon* pulled up on shore, two people in parkas standing near fish-drying racks, and the post buildings in the background. Alaska State Library, Wickersham Collection, Charles Farciot, SL-P277-017—027.

Anglican missionary Robert McDonald impressed Native people with his Native language preaching and literacy projects so that the Merciers' support was not sufficient to establish French-speaking Roman Catholic missionaries there. English was on the rise and Native traders were astute observers of the requirements for successful trade in new circumstances. McDonald, who married into the Gwich'in community, had his brother Kenneth, Bishop Bompas, and the resources of the Church Missionary Society for support. Mercier clearly understood the power of both the messenger and his message, but failed to convince the preacher to settle at his posts.

On the business side, François Mercier had problems with both his Native clients and his nonnative business sponsors and partners. He was from a wealthy and prestigious Québec family, fiercely proud of his heritage. His attitudes may not have impressed the American owners of the A.C.C. in San Francisco, who were of German-speaking background.[358] Mercier was caught up in ethnic difficulties with his Native customers on the Yukon. Shahnyaati' and certain Hän chiefs, such as Bikkinechuti and others, resisted the new American companies he represented, preferring to support their displaced H.B.C. associates.[359] Mercier may have brought prejudices towards Native people from his fur trade experience in the violent American West that did not suit the Yukon. Northern Native traders, used to H.B.C. rituals, prices, and goods, may have balked at his methods, especially if accompanied by the brandishing of six-guns as described in the 1871 Québec newspaper interview with Mercier. Perhaps the destruction the Merciers experienced at Fort Yukon followed threats they made in taking over the fort. The influence of the H.B.C. dogged Mercier's whole career in the North, fading only in the 1880s after the Chilkoot Pass was opened to prospectors, and the Yukon River replaced the Porcupine as the main business scene for Native traders and nonnative customers.

Mercier was proud to employ his French Canadian compatriots as employees, but he was also pragmatic. He saw talent in other men and was quick to sign up the energetic and competent prospectors McQuesten, Mayo, and Harper. Ironically they took over the very territory on the upper Yukon he identified as the next valuable trade frontier. They tenaciously held the area through tough years and many changes in companies and trade practices, empowered by the stability and connections of their Native families, and their partnership as fellow prospectors. They were all English speakers and François Mercier was briefly their sponsor and boss, later an "opposition" company competitor, and perhaps always more a neighbor and acquaintance than friend.

Mercier certainly had friends and acquaintances in both the Native and nonnative communities where he lived. The photograph taken at Fort Mercier in the mid-1880s shows him at the center of his group, surrounded by many people of mixed background, as well as his nephew and his faithful dog Jack.[360] He provided few details of his Native employees in his memoirs and interviews, while other commentators named only Temouski, his Creole interpreter, as his assistant. He had problems dealing with Chief Shahnyaati', who outlasted and outdid all the nonnative traders of his day, ending up with his own supply of goods and business at Fort Yukon. Yet Mercier had sufficient confidence, and presumably the contacts, to set up shop twice in Hän country, first at Fort Reliance with Catsah's help, and again at Belle Isle where he established a post right in the midst of David's Camp. Although he was not the first to live there he obviously felt comfortable moving there for the 1882–1883 season, and it appears to be the location he chose to commemorate his northern life in the Boisseau painting.

Given Paul Kandik's probable Hän origins, he may have been Mercier's entrée to that community. Certainly the routes to the Tanana that Kandik drew on their map identified the area of most interest among traders and prospectors, who knew of the Tanana with its rich furs and possible mineral wealth but needed a platform from which to launch their explorations from the Yukon into the upper reaches of that watershed. Kandik knew the way, with the route anchored on the eastern side at Tthee t'äwdlenn (Eagle Bluff). Mercier was the first to capitalize on that situation by establishing a post there, which was occupied by Arthur Harper soon after he left. So his relationships with Native people, like those with his nonnative acquaintances, included the sharing of information that others would explore and exploit more successfully in future years.

As he left the North for the last time he posed with his colleagues in a "who's who" of pioneer traders and river men, another significant image captured by the Swiss French photographer Charles Farciot in his brief northern career. Mercier was dressed in a formal suit topped by an elegant felt hat such as might be seen on the streets of Montreal, San Francisco, or Paris. He stood head and shoulders above the other "men of the Yukon"—a giant in size, matching his accomplishments over more than a decade of hard work.[361] Leaving at the beginning of the Gold Rush era, he did not meet later travelers and government officials like Ogilvie, McGrath, and Wells, so the stories they heard and recorded from his rivals were of his problems at Fort Yukon and Belle Isle.[362]

As the fur trade was overtaken by the search for minerals, more and more American Anglophones arrived. The McQuesten, Mayo, and Harper trio was better equipped than Mercier, in terms of prospecting interests and knowledge, as well as linguistic and cultural affinity, to serve the new community. Well connected to attract Native fur suppliers, they continued that staple trade through the transitional years until miners could sustain their business. They were celebrated as the early supporters and promoters of the Klondike Gold Rush, with mountains, rivers, and towns named for them.[363] Through their Native wives and children their memory still survives in northern families and communities, at least in Harper's and

This photograph has a typewritten label attached identifying "The White Men of St. Michael and the Yukon River, 1885" as follows: front row (seated on canons) Moses Lorenz, A.S. Frederickson, Dr. W. E. Everette; back row, left to right: John Waldrun, John C. Smith (seated in front of Waldrun), John R. Forbes the Engineer, Arthur Harper, Al Mayo, Captain Chas. Peterson, Joseph La Due, John Franklin, Fred Mercier, Gregory Kokerine. It was taken at Saint Michael in the summer of 1885 as François Mercier prepared to leave the North for the last time. Alaska State Library, Wickersham Collection, Charles Farciot, SL-P277-017—003.

Mayo's cases. Mercier remained a single man, having perhaps a brief liaison at Nulato and one child, but neither his *Recollections* nor any other sources connect him to a northern family. Close ties to his French Canadian roots reduced his need for commitments in the North, but his departures for holidays in the South and prolonged absences may have hampered his business goals. When he left in 1885 the francophone influence along the Yukon dimmed but did not disappear, with Oblate priests reappearing at the end of the decade on the lower Yukon to carry on the missionary work he initiated and encouraged.[364] Prospector Joe Ladue remained to become a founder of the Klondike community on the Canadian side, welcoming a new wave of influential French Canadians at Dawson after 1898.[365]

Mercier was well celebrated when he returned home. As a member of a very influential Québécois family, he associated with the elite of his francophone community, appearing in photographs with his old friend Clut, by then Archbishop of Montreal. He gave interviews to French newspapers and magazines about his northern adventures. In France he presented a lecture to the Geographical Society of Paris, meeting Élisée Reclus, author of a French geographical atlas. Reclus cited Mercier as the source of his writing on the early fur trade and explorations in the Yukon, some of which contained significant errors.[366] François lived another two decades, traveling to Europe and the United States to participate in scientific discussions about the North he knew so well, and joining an expedition to Siberia on behalf of the Geographical Society of Paris. When he died in 1906 he was called "*L'explorateur bien connu*" (the well-known explorer), though that was likely true by then only in his French Canadian community and probably little noticed in English Canada or the United States.[367]

Until recently Mercier's contributions, like Paul Kandik's story, had almost disappeared from memory in the North, especially in English-speaking Native and nonnative communities. For a brief moment—perhaps only for a day or two, these very different men connected somehow in the creation of the Kandik Map. The information they shared with Petroff significantly advanced the mapping of their region, information carried forward through many other maps for decades. Credit for that knowledge went to others, primarily to Ivan Petroff, who published their information without naming them. In the end the memory of the men and their map faded because of the power of the printed word over oral sources, and the dominance of English over both their indigenous Native and pioneer francophone traditions.

5

Mapping the North: Where the Kandik Map Fits In

WHAT IS the significance of the Kandik Map to the cartographic history of the Alaska-Yukon borderlands region and to mapmaking in a worldwide context? At the time it was drawn, it may have contributed some new and important geographic details about the area to outsider understandings of the region. Locally it may have fostered fresh ideas for travel, forged new partnerships, or encouraged old rivalries that led to new trading and prospecting ventures, the establishment of more trading posts, and a growing impetus for comprehensive surveys and cartographic resources. The map is documentary evidence of intensive conversations between Native people and nonnative newcomers concerning the landscape and its resources, a process that had been building for many decades leading up to 1880. It represents the introduction of new opportunities beyond local oral traditions for sharing traditional northern indigenous knowledge, using paper and pencil to create a record, transmitting that information to a much broader audience through publications, and preserving it for future generations in a prestigious though distant library.

The map is entitled "Map of the Upper Yukon, Tananah and Kuskokwim rivers," which indicates the perspective of at least one of the probable participants in its creation—Ivan Petroff. This would have been the logical focus of his request if he was the person who instigated the drawing and lettering supplied by Kandik and Mercier, given the state of geographical information available to him when he prepared for his northern journey in 1880. Other manuscript and published maps predate this map by several decades in detailing parts of these rivers and at least some of the tributaries on each of them. What is unique, and possibly preserved for the first time as a documentary source, is the information relating to trails between these three rivers, including the symbols which likely represent days of travel and stopping points, the Native names for rivers, and the locations of Native villages. That focus is a reflection of the contemporary requirement for new geographic knowledge by nonnative travelers in the Alaska-Yukon borderlands region in order to pursue the next stage in exploring and prospecting, and the participation of Native men like Paul Kandik in the changing economic and social milieu of the region.

In addition to the geographic information revealed, the Kandik Map is significant as evidence of a sophisticated collaborative relationship between Native and nonnative people in creating a map made in the North by northerners, with local knowledge included from a variety of sources. This information was transmitted to Petroff, a southern government official, who subsequently used it to inform a much broader audience outside, advancing the state of knowledge about and interest in the region. Petroff's reports were cited in many subsequent maps and publications, which served as the knowledge base for later explorations and developments, including boundary negotiations and border surveys that made profound differences in the lives of northern people. Rarely did the local sources of information, especially Native experts, gain any substantive recognition for their geographic knowledge in these publications, nor until recent years much input into decisions made about their homeland.

Successive groups of newcomers gleaned as much detail as they could from their southern-based predecessors before venturing north. Certain people like George Davidson, William Healey Dall, and William Ogilvie served as key reference points long after they returned to southern Canada or the United States, continuing to profit from their northern experience and knowledge gained from indigenous residents. Along with fur traders, missionaries, prospectors, surveyors, and other travelers, they gave interviews to newspapers, presented papers at conferences, consulted on government policies, and were cited as authorities when new northern publications and maps were produced.

Unveiling the North[368]

Several published reports with maps were widely available by the 1880s in North America and England, and a number of early southern travelers either carried these maps with them or consulted manuscripts in advance of their travels, then asked northern Native people for directions and assistance in following their maps. The Kandik Map is part of a continuum of sharing geographical knowledge between Native and nonnative people with interests in the North, a tradition that began at the time of earliest intercultural contacts and that continues to the present. Some of the key questions that newcomers asked were about river courses, mountain ranges and passes, trails, food resources, and Native villages, all essential details for the economic and other interests of fur traders, prospectors, missionaries, travelers, surveyors, and government officials. In addition newcomers were concerned about the location of the boundary between Alaska and the Yukon, which started as a line on European diplomatic maps, with coordinates negotiated in a treaty between England and Russia in 1825 for demarcating respective territories, and had no physical presence or consequence for local people at the time. As more outsiders arrived and rivalries between trading companies intensified, the boundary gained increasing importance and its exact location became a matter of great urgency and concern.

Certain key reference points provide markers for analyzing the advances in geographic knowledge about the North by outsiders, starting with Russian explorations at the time of Vitus Bering's trip in 1741. His own famous landmark, Mount Saint Elias, was one of the pivotal points, together with the placement of the 141st meridian from it to the Arctic coast. The arrangement and naming of major rivers, their main tributaries, and larger lakes were also important measures of broadening knowledge and exchange, as well as the location of trading posts and trails. For purposes of tracing these changes the following features are useful as reference points on a series of maps from Bering's time through to the

Klondike Gold Rush: mountains—Saint Elias, Denali, Alaska Range; rivers—the Yukon, Tanana, Porcupine, Pelly, White, Klondike, Kuskokwim; lakes—Frances, Laberge, Wellesley, Birch, Minto; trading posts—Fort Selkirk, Fort Reliance, Belle Isle, Fort Yukon, Fort Mercier, Fort Adams, Fort Nulato, Fort Anvik, Saint Michael, Kuskokwim. By examining the inclusion or absence of these features, the location and relationship to other features, languages used in naming places, and other factors, the sources and patterns of exchange are clarified, lending more context to the representations of landscape and intent of specific maps. Sometimes there is a very direct and obvious transfer of information from one map to another, acknowledged in references printed on the map in some cases; at other times these details are not explicitly linked to the original sources of information.

Al Wright in *Prelude to Bonanza* documented the process by which coastal areas of Northwest North America were explored and drawn with increasing accuracy by European mapmakers.[369] Dee Longenbaugh has documented the role of Russian Alaskan Creoles in detailed mapping and naming of Alaskan coastal and interior areas exploited by the Russian-American Company.[370] Early fanciful drawings of the arctic regions by European cartographers speculated on the relationship between Asia and North America[371] and the true outline of the landforms was only gradually "unveiled" as Europeans pushed east from Russia, west from Britain, and north from the United States and Spanish possessions. The search for a Northwest Passage from Europe to the splendid opportunities of Asian markets, together with imperial ambitions and the promise of fabulous riches from northern fur resources, fueled this drive into the Northwest. On the western Pacific side, Russian expeditions from Vitus Bering in 1741 onwards to Zagoskin, and others through the late Russian American period, made steady progress in mapping the Aleutian islands and the coast of Alaska incorporating local knowledge, which was then available primarily within the confines of the Russian-American Company and Imperial Court. The expeditions of British explorers Cook in 1778 and Vancouver in 1792–1794 provided detailed English charts of the coast as far north as Norton Sound. A few Spanish and French explorers also traveled north along the coast for some distance, producing charts in their languages for their sponsors, contributing to a growing interest and exchange among Europeans about the routes and trade opportunities between northwestern

North America, Asia, and Europe. Expedition sponsors usually regarded geographical information as a proprietary resource but captains, crews, and navigators met and talked about their travels in ports around the world. Also, Delisle and other geographers conducted their research and inquiries, gleaning information from the returning adventurers and producing maps that were sketched and discussed around the ports, courts, and salons of Europe. While literacy was a skill confined to people in the elite business and religious circles, a growing number of publishers printed broadsheets, books, maps, and atlases, providing a ready venue for new discoveries and serving eager audiences who circulated the news within and across national boundaries.

British Captain James Cook described the Yukon River delta in his reports while searching for the Northwest Passage. He surmised there was "a considerable river" flowing from the east into Norton Sound, noting that "the sea appeared checkered with shoals—the water very much discolored and muddy, and considerably fresher than at any of the places we had lately anchored."[372] A map published in London in 1780 by C. Cooke[373]

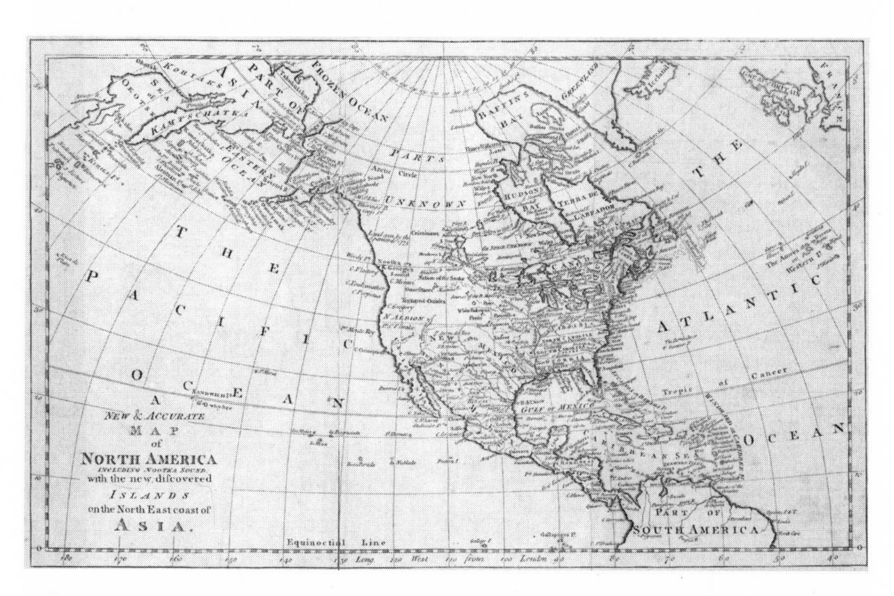

This map was titled *A New & Accurate Map of North America including Nootka Sound: with the new discovered Islands on the North East Coast of A4sia*, published in London c. 1780 by C. Cooke. The interior of Alaska and the Yukon are labeled as "Parts Unknown" as the map predates the northern explorations of Alexander Mackenzie and later Hudson's Bay Company traders. University of Alaska Fairbanks, Rasmuson Library, Alaska & Polar Regions Collections, Rare Map G3300 [1780] C66.

captures the significant information learned from the earliest Russian and English explorations in the Northwest. Showing the coastal outline of all of North America, the eastern Asian coast of Russia, and a small part of northern Europe, the whole interior northwestern section of North America is labeled "Parts Unknown," and the Arctic Ocean is labeled "Frozen Ocean," with no detail illustrated west of Baffin's Bay in northern Canada or east of Icy Cape in Alaska. On the southeast coast of Alaska, "Behring's Bay" is designated, as is Mount Saint Elias, but with no boundary line drawn to the north, as this predates the Anglo-Russian treaty of 1825. Norton Sound is shown and named on the northwestern coast with a notation of "Shallow Water" at the location of the Yukon River delta, but no indication of a river course into the interior.

By the time *Tebenkov's Atlas* was published in 1852,[374] the combined knowledge of Russian Alaskan Creoles and R.A.C. officers, British and other explorers contributed to detailed drawings of coastal Alaska and the lower sections of the great rivers of the interior. Russians and Creoles had explored and mapped along the lower Kuskokwim and on the Yukon, which they called the Kwikpak,[375] upstream to Nowitna (near present-day Ruby), only a hundred miles short of the Tanana.[376] A Russian map dated 1861 names hundreds of places along the Alaskan coastline and depicts the complexities of the Yukon delta and its course upriver well past Nulato, with Mount Saint Elias prominently marked on the southeast coast and the 1825 border stretching north to the Arctic coastline.[377] The upper reaches of the Yukon on the British side of the boundary are sketched in vague lines, probably including information gleaned from published maps of British North America. There is a blank space in the middle between the British section of the Yukon and the Russian Kwikpak, reflecting the lack of direct travel and observation of this area by R.A.C. officers to that date. It could have been charted using the reports and sketches of Creoles and interior Native traders who went up and down the river; however, the Russians may have preferred to record only the details of areas actually traveled by their own employees. The following year this section could have been verified using information from Ivan Lukeen's pioneer trip from Saint Michael up to Fort Yukon, which dispelled any doubts about the Yukon and Kwikpak being one and the same long river.[378]

This Russian chart bears the undated signature of George Davidson, and it may be one of the charts he viewed in 1867 when he visited the headquarters of the Russian-American Company in Sitka as an officer of the U.S. Coast and Geodetic Survey, shortly after the U.S. purchase of Alaska, or it may have been a chart he acquired later for his own collection. His signature and another map published by Colton & Co. of New York shortly after the Alaska Purchase clearly demonstrate the passage of Russian geographic knowledge to the new American owners of the territory. The Colton Map again shows Mount Saint Elias with the boundary line to the Arctic at the 141st meridian and the whole course of the Yukon River named as such, as well as the Porcupine, White, Pelly, and Lewes (the Yukon above the Pelly confluence) rivers. The Tanana is not named but is shown flowing to the west from the Yukon confluence, reflecting a mistaken view of its course that persisted for many years. A river, which is probably the Tatonduc, is shown flowing north to the Porcupine,

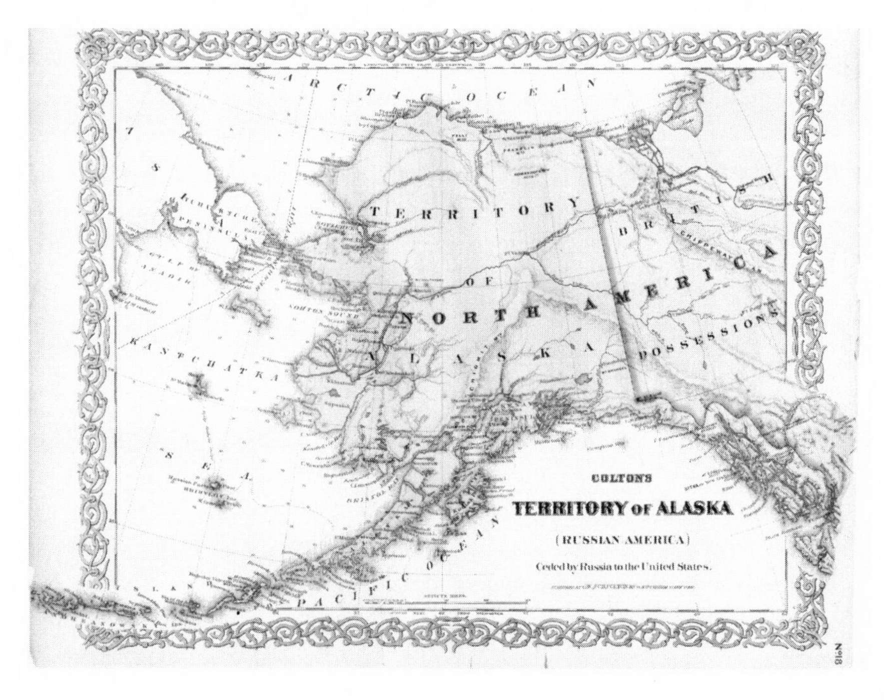

This map titled *Colton's Territory of Alaska (Russian America) Ceded by Russia to the United States* shows the full run of the Yukon River from its headwaters on the Lewes to the Bering Sea. The Pelly and Porcupine are named, but not the Tanana, which is depicted flowing from the southwest into the Yukon below Fort Yukon. The map postdates the sale of Alaska to the United States in 1867. University of Alaska Fairbanks, Rasmuson Library, Alaska & Polar Regions Collections, Rare Map G4370 [1862?] C631.

but not named. Several H.B.C. posts are located, including Fort Yukon clearly on the American side of the border, Fort Selkirk, Fort Pelly Banks, Fort McPherson, and Lapierre House, information probably obtained from British or Canadian maps showing H.B.C. posts in the Northwest. Although this map is tentatively identified as being published in 1862 in the University of Alaska Fairbanks Archives Map Collection, it must have originated sometime after 1867, since it has the subtitle "Russian America Ceded by Russia to the United States." It probably was published before 1869 when George Davidson's U.S. Coast and Geodetic Survey map of Alaska plus the first *Alaska Coast Pilot* were available, both including more details of the Yukon above Fort Selkirk derived from the exploration maps of the Western Union Telegraph survey in 1866, as well as information from Tlingit Chief Kohklux' maps.[379]

The eastern side of the Alaska-Yukon region was explored and mapped by Canadian fur traders and British explorers with their Native guides pushing west and north across the continent from Canada. In 1789 Alexander Mackenzie completed the first voyage north to the Arctic. Mackenzie's journals were known to the Russians and there was much subsequent speculation about the origins and course of a "great river" described by the Native people, who were undoubtedly Gwich'in, whom he met somewhere around the Peel River. Mackenzie described a drawing of the middle stretches of the Yukon River, sketched by one of these people in the sand, showing "a very long point of land between the rivers . . . which he represented as running into a great lake, at the extremity of which, as he had been told by Indians of other nations, there was a Belhoullay Couin, or White Man's Fort. This I took to be Unalascha Fort, and . . . the river to the West to be Cook's river; and that the body of water or sea into which this discharges itself . . . communicates with Norton Sound."[380]

The steady westward push of H.B.C. traders was under the direction of Governor George Simpson, a veteran of Canadian exploration and keen student of the growing body of mapped knowledge of the Northwest. He developed a plan for two routes of exploration across the continental divide from the Mackenzie Valley, starting with John Bell's 1842 trip that located the Porcupine River on H.B.C. maps, and his 1844 voyage to its mouth, where he met Indians who called the larger river at the confluence "the Youcon," a name he adopted and relayed back to his supervisors.[381] Simpson speculated that this westward flowing river was the Kwikpak, known to the Russians since various Indians told Bell about other white traders not far downstream. Murray's establishment of Fort Yukon in 1847 was well known to be in Russian territory so that some H.B.C. maps of the time either neglect to show the post or are vague as to the location of the boundary at the 141st meridian. Simpson sent Robert Campbell to explore the upper reaches of the Liard searching for a more southerly route to the "great river," which added the Liard, Frances Lake, and the upper Pelly to the H.B.C. map by 1843. Then in 1847 Campbell followed the Pelly to its confluence with another large river he named the Lewes flowing from the south, building Fort Selkirk there in 1848, initially on the south bank of the Pelly. Later he moved the post to the west bank of the Pelly-Lewes, the site of Fort Selkirk today. With two new posts tenuously established west of the Rocky Mountains, Simpson gave the order for Campbell to travel downriver from Fort Selkirk in 1851 to verify what the shrewd Governor already suspected—that the Pelly-Lewes and the "Youcon" were one and the same, and very likely the same river that flowed to the Russians' Kwikpak and the Pacific. When Campbell paddled downriver to Fort Yukon in 1851, he named the Stewart and the White rivers en route, as well as several other tributaries.[382]

After the Tlingits ousted Campbell from Fort Selkirk in 1852, he traveled home to Scotland for a long-delayed furlough, stopping in London to share his knowledge with John Arrowsmith, official cartographer of the H.B.C. The results can be seen in the 1854 Arrowsmith *Map of British North America*, the first published reference that traced the Yukon from the southern stretches above Fort Selkirk all the way to the Pacific Ocean, with the sections from the confluence of the Porcupine down to an unnamed tributary in the vicinity of the Nowikat appearing as a double set of dashed lines, indicating that area was not definitively known to the H.B.C.[383] Although this map predated Lukeen's trip of 1861, the H.B.C. men at Fort Yukon had plenty of evidence in the form of trade goods, stories, and sketches from their Native customers to conclude that the Yukon flowed to the Pacific. While not credited to Native explorers and mappers, this Arrowsmith Map clearly owed a debt to them. The Yukon River is not named anywhere on this map as such, with the stretches above the Porcupine designated as the Pelly River, and only the mouth labeled as the "R. Kvichpak." This reflects the prevailing opinion that the headwaters of the Yukon were in fact on the upper Pelly, and the reason for calling the river above Fort Selkirk the Lewes, the name attached to it on this map, which continued to be used for many decades to come.[384] Included on the same map are the H.B.C. names for tributaries between Fort Selkirk, which is clearly marked, and the Porcupine River, as well as the 1825 boundary at the 141st meridian.

Fort Yukon is not named on the map, where it would have been located very obviously well to the west of British territory. The map is inscribed "By permission dedicated to the Hon.ble Hudson's Bay Company; containing the latest information which their documents furnish."[385] Clearly not all of the "latest information" was deemed appropriate for publication at this time! The Tanana is named "Mountain Men R.", shown as a set of dashed lines leading south from the vicinity of Nuklukayet (not named) and hinting at the activities of the H.B.C. in the area, despite the obfuscation of their post-building activities at the mouth of the Porcupine. Ironically the map does include information about the upper Yukon that Campbell obtained from the Tlingits who ousted him. The map shows the river south of Fort Selkirk all the way to Lake Laberge, though not in much detail.

From the 1840s through the 1860s several H.B.C. men at Fort Yukon gathered geographic information from their Native contacts to fill in the blanks on their maps of the interior, especially related to the Tanana, the source of a huge array of fine-quality furs. Murray produced a sketch map, which has not survived, but many of the details were reproduced in a map in Sir John Richardson's *Report* of 1857.[386] Murray's notes offer a fascinating picture of how he pieced together his information: "I have seen two Indians who were at the Fort on the coast and acquainted it with the inland route, I had them describe it to me and chalk it down on the floor. The river they ascend from the coast must as far as I can judge fall into Norton Sound, or perhaps Kotzebue Sound, but I think the former as there were two large vessels at anchor while the Indians were there."[387] These same Indian men described another great river that flowed to the Pacific, which Murray surmised was the Kuskokwim. Before Campbell made his famous trip from Fort Selkirk to "confirm" the connection between the Pelly and the Yukon, Murray had already verified the matter with his Native mappers: "In the first place the Pelly and the Yukon are one and the same. Two Indians of the upper 'Gens du fou' who had been at the Pelly were here in summer, and with them another Indian belonging to the 'Men of

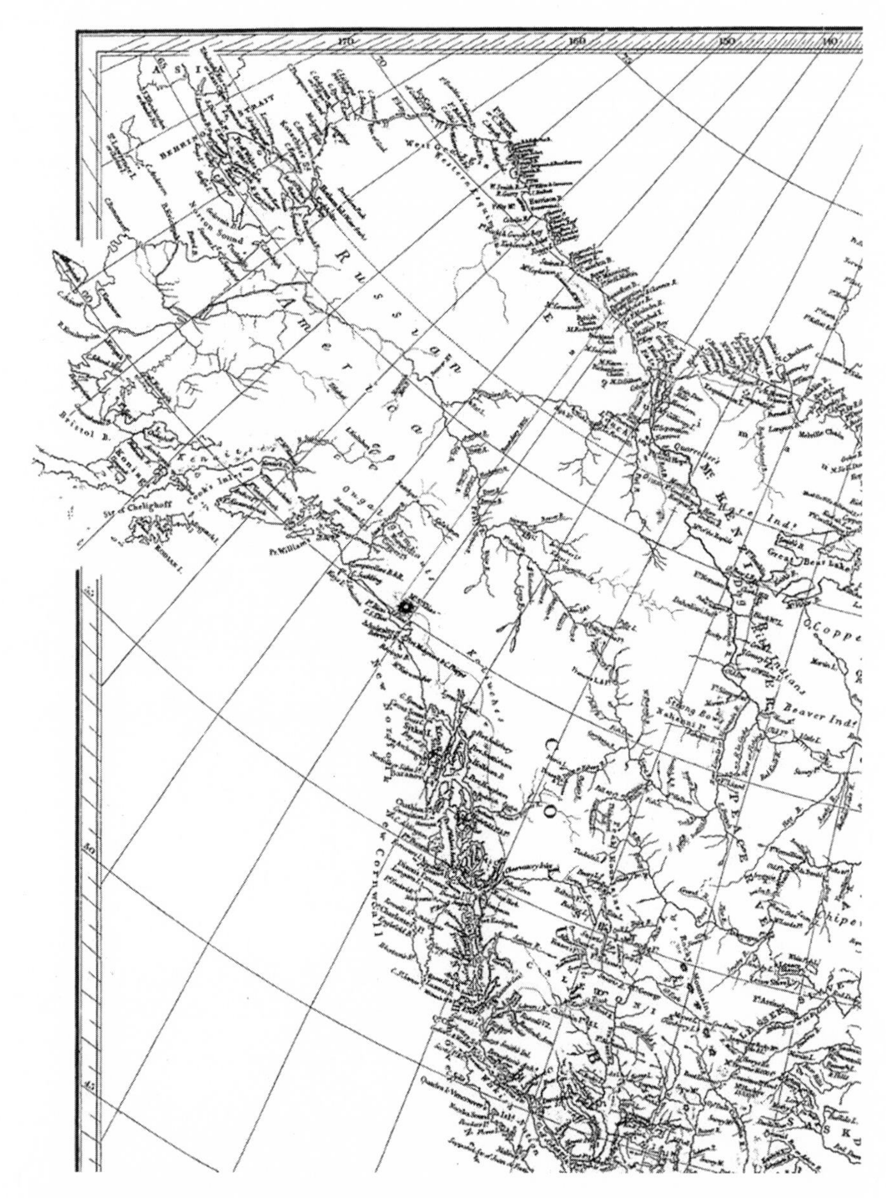

This map produced by the Arrowsmith firm of London, England, in 1854 published the first detailed drawing and naming of rivers and places in the interior Yukon. H.B.C. trader Robert Campbell brought the new information to the cartographers, including details he obtained from coastal Tlingit people, interior Athabaskan people, and explorations by himself and some of his H.B.C. colleagues. J. Arrowsmith, *British North America 1854 Reproduction*. Ottawa and Whitehorse: Association of Canadian Map Libraries and Yukon Historical and Museums Association, 1982.

the Forks' (a band near to the forks of the Lewis and Pelly) who had two years before been at the Great Lake, the principal source of the river."[388] Murray used their information and some of Campbell's sketches he saw at Fort Simpson to fix the location of Frances Lake and the confluence of the Pelly and Lewes. He knew that the course of the Yukon from the Pelly "as drawn by the indians [*sic*] is to the north west, and at one place passes between high rocks or ramparts. . . .The next river of any importance is 'Red Island River' which joins it from the north west." Below Fort Yukon, the river was joined "by the 'River of the Mountain Men' [that enters] from the south and runs nearly parallel with the Youcon. . . . Next comes what I have marked as Russian River . . . [where] the Russians have wintered and are now established. Below that very little is known by the Indians here."[389]

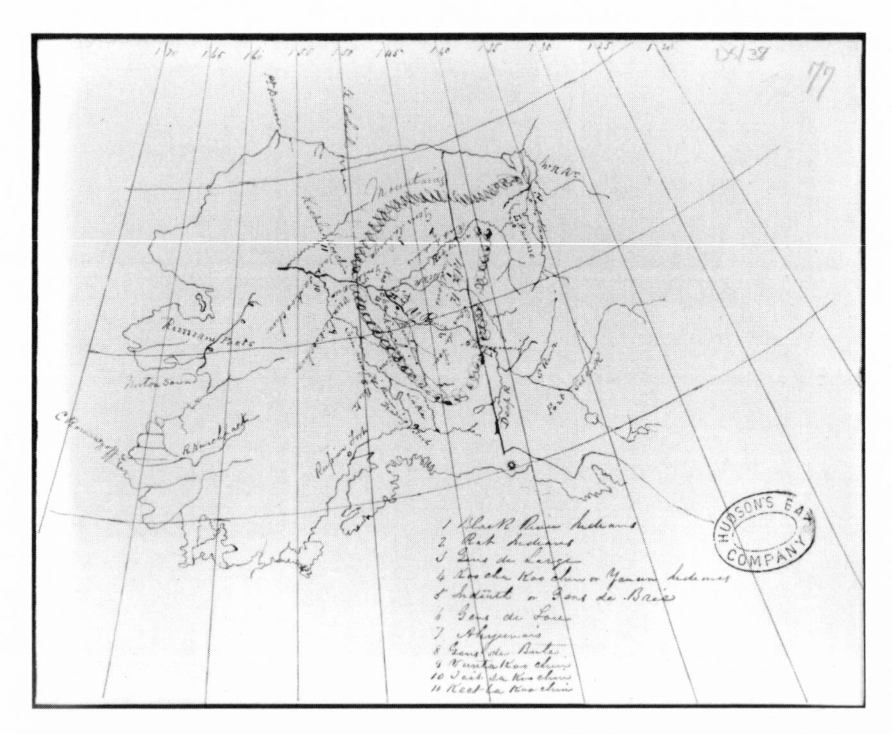

William Hardisty, an H.B.C. trader at Fort Yukon for several years, drew this 1853 map of the areas utilized by different groups of hunters and trappers in the region surrounding the post. The Yukon, Tanana, and Porcupine rivers are shown, with a gap between the upper Yukon and the lower river, which is labeled with its Russian name written as "Kwichpacta." Hudson's Bay Company Archives, Archives of Manitoba, HBCA D.5/38 fo. 77.

Two more maps demonstrate the considerable geographic detail the H.B.C. men gleaned from various Native groups trading at Fort Yukon. William Lucas Hardisty drew a sketch map in 1853 locating the Yukon, Porcupine, Tanana, and Kuskokwim, together with the locations of the different groups of people in the region.[390] The Kwikpak and the upper Yukon are not connected and the upper Kuskokwim region is a blank space on his map.

In 1872 James McDougall drew another map, again locating the various groups of Native people and showing the relationship between the Yukon, Porcupine, and Tanana, including a number of tributaries on all three, though only a few are labeled with names.[391] McDougall was the unfortunate and disbelieving recipient of U.S. Army Captain Charles Raymond's orders to vacate Fort Yukon in 1869, hoping for some time that the orders might somehow be rescinded. Although Raymond reported

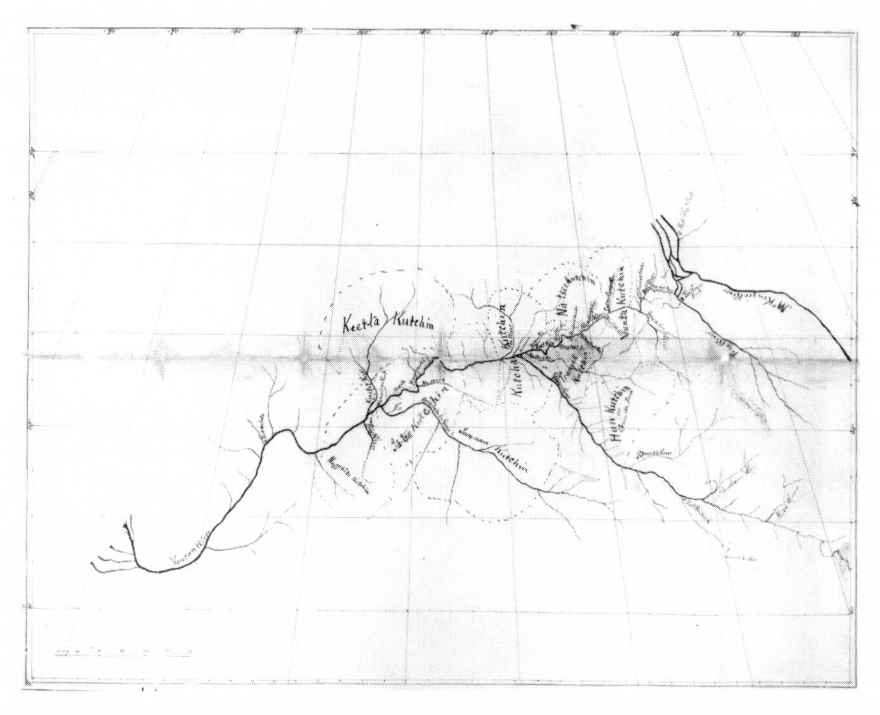

James McDougall was another H.B.C. trader who lived at Fort Yukon for many years. He drew this sketch map of the Yukon River in 1872 showing the names and territories of various groups of Native people in Alaska and the Yukon, H.B.C. trading posts, and tributaries of the Yukon, Tanana, and Porcupine rivers. Hudson's Bay Company Archives, Archives of Manitoba, HBCA E.38/25 fo. 54.

that he received the utmost cooperation and courtesy from the H.B.C. men at Fort Yukon,[392] it appears that McDougall's hospitality might not have extended to sharing the wealth of data collected in past years from the local people with any of his American visitors in the 1860s. Neither Dall's map, made after his trip up the Yukon in 1866 and published in 1870,[393] nor Raymond's survey of 1869,[394] which was published in 1871, showed much of the Tanana. They included a short stretch of the river roughly as far as Lake Minto, but neither had as much detail as McDougall apparently knew at the time.

Defining the Boundary

With the takeover of Alaska in 1867, various U.S. agencies began to collect data on the geography and resources of the new American territory. As all commercial activity depended on maritime shipping, the immediate focus was on producing reliable charts translated from Russian into English, verified and amplified with new data, for American ships sailing in Alaska's coastal waters. George Davidson was assigned to head up this task and traveled north in 1867 and again in 1869 to assemble data, consult Russian sources, and conduct the surveys needed to complete the *Alaska Coast Pilot* in 1869.[395] He was also assigned to travel to Klukwan to observe the solar eclipse of that year, coordinating his observations with those of Captain Raymond at Fort Yukon to verify the longitudinal calculations that would help define the location of the boundary at the 141st meridian. The events surrounding the observation of the eclipse inspired one of the most remarkable exchanges of geographical information between Native and nonnative people recorded in northern history, resulting in the production of the Kohklux Map of Tlingit routes to the interior Yukon.[396]

Kohklux was the son of the great Klukwan Chief Skeet-lah-ka, who led the 1852 raid on Campbell at Fort Selkirk. Both father and son had traded with the Russians at Sitka, and H.B.C. and American trading ships at Pyramid Harbor in the late 1840s and 1850s.[397] Kohklux also traveled with his father on annual trading trips into the interior Yukon. The map he and his wives drew for George Davidson in 1869 showed the upper reaches of the Yukon River, with connecting tributaries, lakes, and trails

In 1869 the Tlingit Chief Kohklux and his two wives drew this map of their trade routes to the Yukon interior at Klukwan, Alaska, for American scientist George Davidson who was there to observe a solar eclipse. The map was drawn on the back of a large coastal chart and took three days to complete. Davidson annotated the map with the Tlingit place-names and other travel details provided by the mapmakers. Kohklux, Map of Tlingit Trade Routes, The Bancroft Library, Map Collection, G4370 1852 K6 Case XD.

all the way to the White River on the north, with the Pelly and Frances Lake on the upper right-hand corner, the Teslin at the southeast, and the Alsek and Tatshenshini on the southwest. The most intensive detail was sketched for the areas closest to Klukwan, including individual mountain peaks, glaciers, rivers, portages, and days of travel going to and returning from Fort Selkirk, as well as portions of the Lynn Canal. George Davidson wrote Tlingit and Tutchone names for important landmarks on the map, following the instructions of his Tlingit hosts.[398] Davidson returned to his office at the U.S. Coast and Geodetic Survey in San Francisco where he used this map to inform his knowledge of the Lynn Canal, the coastal passes, and routes to the interior Yukon, consulting it in the following decades and showing it to other government officials, explorers, writers, and mapmakers interested in Alaska-Yukon,[399] possibly including Bancroft and Petroff. Although the Tlingit contributions were not credited on his official charts during these years, Davidson wrote an article for *Mazama Magazine* in 1901 that told the story of Kohklux and his wives drawing the map, linking the information they shared in 1869 with the travels of later white explorers.[400]

While nonnative people were keen to learn all they could from indigenous travelers, Native people were also curious about the methods used by the newcomers to observe and record geographic information. Davidson reported that the Tlingit people were astonished by his ability to predict the timing of the 1869 eclipse and his ability to write and then repeat back to them the Tlingit names for rivers and other features on their map.[401] Raymond recorded his experience with Gwich'in people watching his work at Fort Yukon: "Some of them were very much interested in my operations, and I found no difficulty in making them comprehend, although through an interpreter, the general method and purpose of my astronomical observations. Indeed they are accustomed to note time roughly by the relative positions of the stars."[402]

Raymond's map was the first comprehensive and official chart published by the United States showing the interior of Alaska after its purchase in 1867. Entitled "The Yukon River, Alaska from Fort Yukon to the Sea," it was issued in 1871 by the Office of the Chief of Engineers, War Department, reflecting the serious side of charting, naming, and claiming the new territory. The map included Raymond's detailed rendering of the course of the Yukon from the Pacific to Fort Yukon, many large and small tributaries, and the topography on both sides of the river as sketched from the boats along the way, together with names of places supplied by local informants, information obtained from George Davidson and William Healey Dall, as well as some from the H.B.C. men. His survey and resulting map marked the first direct government intervention on the ground to bring into effect the boundary originally negotiated by Britain and Russia in 1825 that had up until then remained an imaginary line with no ground truthing. Raymond himself did not actually travel far enough up either the Yukon or the Porcupine to mark the boundary, but he used his longitudinal reckonings to provide the H.B.C. men with an estimate of how far upriver they would have to move to build a new post on the British side of the border.

The other pressing issue for the new American enterprises on the Yukon was navigating the shifting channels and swift waters of the Yukon. On board the Parrott & Company steamer with Raymond and the traders in 1869 was an unnamed Native guide acquainted with the river, no doubt a fascinating first voyage for him, as he transferred knowledge and skills learned while paddling a canoe to this new and very different form of river travel.[403] Raymond included an insert map of the whole of "Alaska and Adjacent Territory" on his published chart, which shows the boundary line along the 141st meridian and down the panhandle, the coastal outline of Alaska, the Kuskokwim to a similar extent as the last Russian charts, Saint Michael, Nulato, Fort Yukon, and Fort Selkirk. The name Yukon River is located along the stretch between Nulato and Fort Yukon and the whole river is shown from the mouth to the upper reaches of the Pelly (not named) and the Lewis. The upper stretches of the river appear to include some of the detail shared by Kohklux with Davidson, as it is a more detailed sketch of the upper lakes and rivers than appeared on the H.B.C. Arrowsmith maps of the 1850s, a reasonable surmise since Raymond and Davidson communicated with each other before and after their respective travels to Alaska in 1869 to coordinate their observations of the eclipse and other data.[404] The Chilkat River and the "Tahco" River are shown with a series of connecting lakes and streams leading to the Lewis, though not anything like the detail Kohklux provided on his map. Raymond's insert resembles the Arrowsmith maps of 1854 and 1857 in its depictions of the shapes and configurations of Frances Lake and the Pelly River, and the naming of downstream tributaries to Fort Yukon conforms to names supplied by Campbell and Murray in the 1850s, so he could have had access to those published maps, but probably not the more proprietary information on the Tanana located on the H.B.C. manuscript maps produced by the Fort Yukon H.B.C. men.

William Healey Dall published his report together with a map in 1870, with essentially the same information and vision as Raymond of the lower Yukon River, but including a more detailed drawing and names for the upper Yukon. He was the first to name Lake Laberge,[405] honoring his old friend Michel Laberge, who had heard of the large lake at the upper Lewes from the Indians at Fort Selkirk in 1866 but did not actually travel there to "discover" it.[406] Since Laberge was a French Canadian and the lake was clearly in British territory, Dall perhaps was exercising some diplomatic discretion in his choice of the name, which continues in use to this day. It had a Native name in the local Southern Tutchone language—"Ta'an Mün"[407]—and also a Tlingit name, "Kluk-tak-syee," as shown on the Kohklux Map, but few nonnative explorers or mappers of the time thought to retain Native place-names. Dall's report and his map showing

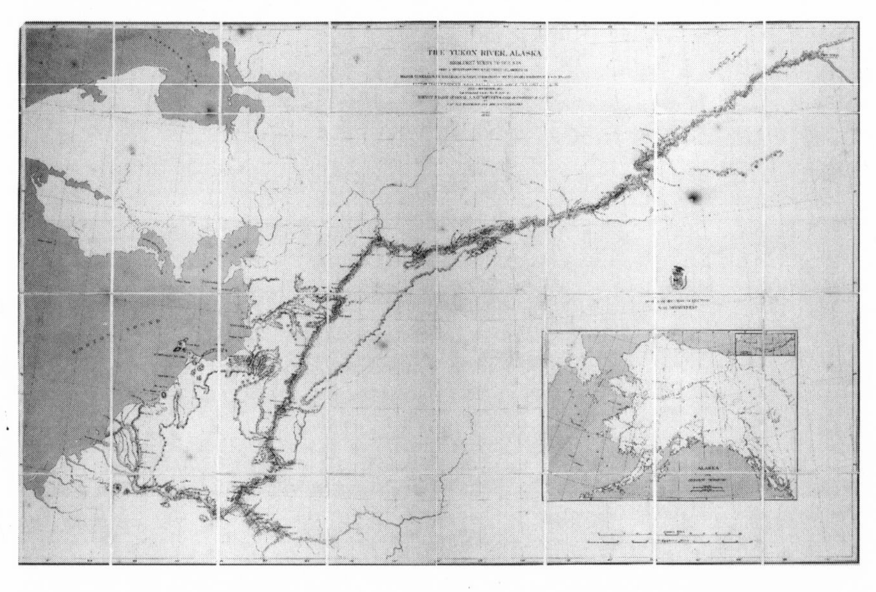

This detail from Captain Charles Raymond's map published with his report on his Yukon River trip in 1871 shows the limited information he obtained regarding the Tanana River in 1869. Captain Charles Raymond, *The Yukon River*, Alaska from Fort Yukon to the sea. University of Alaska Fairbanks, Rasmuson Library, Alaska & Polar Regions Collections, Rare Map digital ID f87101.

the Yukon River from its headwaters south of Lake Laberge all the way down to the Pacific represents the knowledge accumulated by then through direct explorations by nonnative traders and travelers, plus information gleaned from Native people they met who traveled beyond the areas then utilized by nonnatives. This pattern of continuous sharing of knowledge between Native people and newcomers of various backgrounds persisted from the 1860s through the 1870s, though little new information appeared on maps published in that time, with later editions of Dall's map repeating the information published initially with his report.

With a number of new publications, more people traveling to the region had read or heard about various geographic features or had maps in their possession when they arrived. Whymper's book, published in 1873, included geographic descriptions and a map of his travels with Dall in the 1860s. Their sponsor, the Western Union Telegraph Company, also

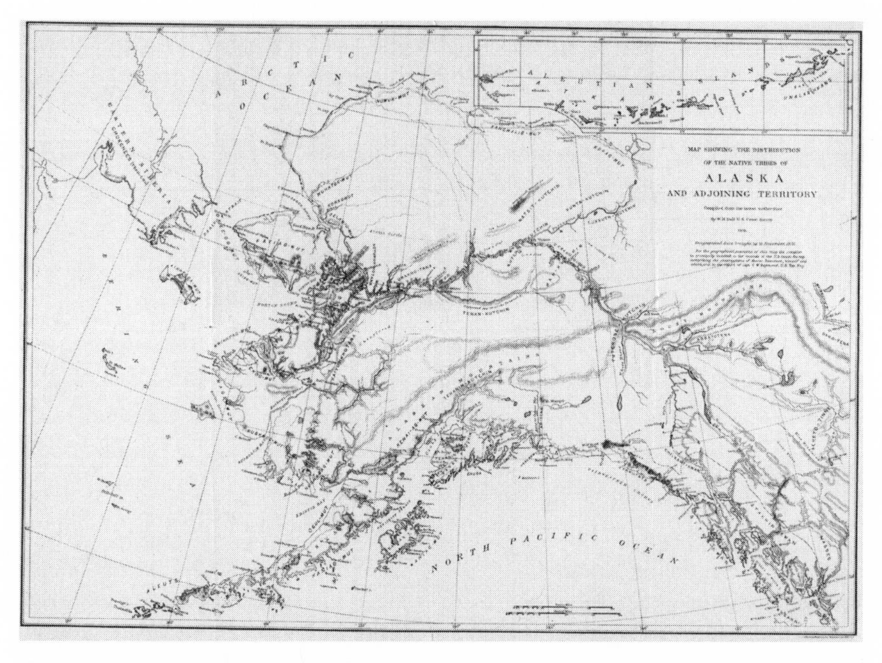

On his 1875 *Map Showing the Distribution of the Native Tribes of Alaska and Adjoining Territory* William H. Dall cited information he received from George Davidson and Charles Raymond plus his own observations as sources for the updated geographical information he portrayed. His map depicts the full course of the Yukon River but only a short section of the Tanana with the notation that this is the "supposed course" of the river. University of Alaska Fairbanks, Rasmuson Library, Alaska & Polar Regions Collections, Rare Map G4371.E1 1875 D3.

produced maps of the proposed route through the Northwest based on reports of their travels.[408] Whymper noted that Ketchum and Laberge's depiction of the upper Yukon as far as Fort Selkirk closely resembled an Arrowsmith Map he saw in England. Robert McDonald gave a copy of some unspecified published map of British North America to the A.C.C. Captain Riedelle in the early 1870s.[409] McDonald was also one of the first "discoverers" of gold, shown to him by his Native converts in the vicinity of Preacher Creek and reported in an article in the *London Times* during the 1860s.[410] McDonald and other missionaries like Bishop Bompas were constantly seeking out the areas and names for places where people camped in order to preach to them, and then sending the information back to the Church Missionary Society in England.[411]

Gold was a powerful attraction and motivator for travelers inclined to prospecting. Arthur Harper had a copy of an *Arrowsmith Map of British North America*, which convinced him there was gold in the Yukon watershed.[412] When he arrived at Fort Yukon in 1874 Harper "met an Indian who had quite a chunk of native copper. Anxious inquiry elicited the response that it came from the White River, more than four hundred miles up the Yukon."[413] Harper immediately left to explore the White River, though no other traders had been there previously and no maps existed to guide the way. He did have Native guides who also helped to hunt and gather provisions for surviving the winter. The next year he briefly considered a trip up the Fortymile after finding some good prospects at the mouth of the river, but "some Indians camped there made them believe that there was a terribly dangerous canon some distance up."[414] So Harper and his partners continued back upriver to spend a second winter on the White. Ogilvie's words written years later illustrate the misunderstandings that occurred at times between Native and nonnative people when different perspectives and purposes clashed: "We may pause here and ask, if they had known then, what they became well acquainted with not long after, that the Indian description was a gross exaggeration, and had gone up the Fortymile. . . . What a difference in history if they had discovered the gold later found by Franklin and Madison."[415] In fact tragedy struck a Tanana family on the Fortymile shortly after the first gold discoveries there. The McQuesten, Mayo, and Harper consortium set up a post at the mouth of the river and apparently closed the Belle Isle post at Mission Creek so the Tanana people were coming down the Fortymile to trade in the spring. "The Black Showman" (or shaman), as the traders knew him, capsized in the rough waters of the Fortymile and saw his wife and children swept downriver in the torrents. In despair he flung himself into the icy waters and drowned, not knowing that miners downstream had rescued his family. This story was told repeatedly to Ogilvie and other newcomers, and inscribed on a photo of the Tanana man taken by Charles Farciot the previous spring.

The tragic story of this man known as "the Black Shaman" circulated through all the communities of the upper Yukon and Tanana valleys in the mid-1880s. His canoe overturned in the turbulent waters of the Fortymile River. He managed to reach shore but jumped back into the frigid waters and drowned after seeing his family swept away. Some boatmen saved his wife and children downstream. Alaska State Library, Wickersham Collection, Charles Farciot, SL-P277-017-015.

Native people frequently provided samples of various ores of interest to the trader-prospectors. Harper was prospecting in the vicinity of the mouth of the Tanana in 1874 and was shown some small nuggets said to be found nearby but was unsuccessful and moved to locations upriver on the Yukon and the White River, where Indian hunters reported copper nuggets.[416] In 1875 he traveled from Fort Reliance up to Mission Creek and from there crossed to the north fork of the Fortymile, followed it to the main river, then crossed the divide to the Sixtymile, "where he found good pay" but was never able to locate the area again.[417] According to Ogilvie, "some time after this he made an exploratory trip to the head of the Tanana, going up the Fortymile to reach it. He found good prospects but could never afford to give up his business with the company and take to mining exclusively."[418] This may have been the 1881 trip with Bates. Another A.C.C. employee, George Holt, sent some gold out to Saint Michael in 1880 that was given to him by a Tanana River Indian.[419]

These stories of gold and copper in the Tanana and Fortymile regions were circulated by Harper and others in letters to friends in the South so that interest was building in the possibilities of new strikes as the mining boom in the Cassiar and other coastal districts dissipated. By the time of Petroff's arrival in 1880 to take the Alaska Census, the upper Yukon, Tanana, and Kuskokwim were ripe for exploration and documentation. Petroff claimed to have conducted "researches in public and department libraries, archives, etc." and consulted "all the ancient and modern maps and charts (Russian, English, French, and American) accessible to me." [420] so he was likely aware of the gaps in knowledge related to interior river drainages. When all the river traders assembled at Saint Michael that summer there would have been many discussions about who had the most furs and speculation about who had been prospecting in which areas and who intended to go on to new areas next.

The map of Alaska included with Petroff's first preliminary report on the Census, hastily published in 1881, showed the 141st meridian to the west of Fort Reliance, placing it in British territory. In the subsequent longer *Report* published in 1882 Petroff devoted a whole section to boundary issues left murky from the days of the 1825 treaty between Russia and Britain. He anticipated problems if the issues were not resolved in the near future and was careful in what he showed on his maps. Advancing the interests of his U.S. Government employers, he proposed solutions much to the advantage of the American traders then operating on the upper Yukon: "A survey . . . would be altogether too costly, but a straight line between certain easily defined points agreed upon by mutual consent would solve a difficulty which promises to arise in the near future, owing to the discovery of valuable mineral deposits on the very ground placed in dispute or doubt by the old treaty. It may be stated here that a line drawn . . . [from near present-day Hyder, the eastern extremity of southeast Alaska] . . . to the intersection of the sixty-fifth parallel with the one hundred and forty first meridian would nearly follow the present line in southeastern Alaska, while it would give to the United States one of the head branches of the Yukon river—the main artery of the trade of the continental portion of Alaska—which is now crossed by the boundary at a point considerably below the head of steam navigation."[421] The maps published with the 1882 *Report* are significantly different from the 1881 version, showing Fort Reliance just to the west of the 141st meridian in American territory, an interesting echo of the earlier days of the H.B.C. concerns about the location of the line. It is difficult to assess whether the information supplied to Petroff by the American traders about the supposed location of the boundary was "genuine" wishful thinking or "ingenious," like the earlier British fictions about Fort Yukon, or his own inventions. Further research among archival correspondence in U.S. repositories might reveal more details. Petroff's musings about new ways of drawing the 141st line and hints at problems over future gold prospects point to possible self-interest and potential national gain in envisioning the line to be on "the right side" of U.S. business interests.

Throughout the report Petroff made numerous references to the geographic knowledge and contributions of Native people. He noted that the watershed between the Chilkat and the Yukon had been drawn using the recent discoveries of the German explorer Aurel Krause, who had been taken through the mountain passes by Klukwan Tlingits, including one of Kohklux's sons.[422] Tlingit people provided Petroff with "minute descriptions" for Dry Bay. In the interior, "the course of the Kuskokwim river has been retained from [the 1869 coast survey map] with the exception of a portion of its headwaters corrected from Indian maps and the descriptions of traders."[423] He relied on Captain Raymond's map of 1869 for the course of the Yukon River, noting that Raymond had found Fort Yukon to be located considerably west of previous mapped locations. "For the course of the river between fort Yukon and the British boundary I am indebted to magnetic bearings furnished by traders traveling on the steamer which ascends the Yukon to fort Reliance, an American trading-station. Those bearings, confirmed by Indian maps and the descriptions of various intelligent individuals, when brought into connection with the change in the position of Fort Yukon bring Fort Reliance within our possessions, though heretofor [*sic*] it was supposed to be on British territory, owing to deductions made from the erroneous location of fort Yukon. The course of the Tennanah river and that of the portage routes connecting this little-known stream with the Yukon on the east and the Kuskokwim on the west are represented in accordance with Indian maps and a careful comparison of statements of many traders and intelligent natives."[424]

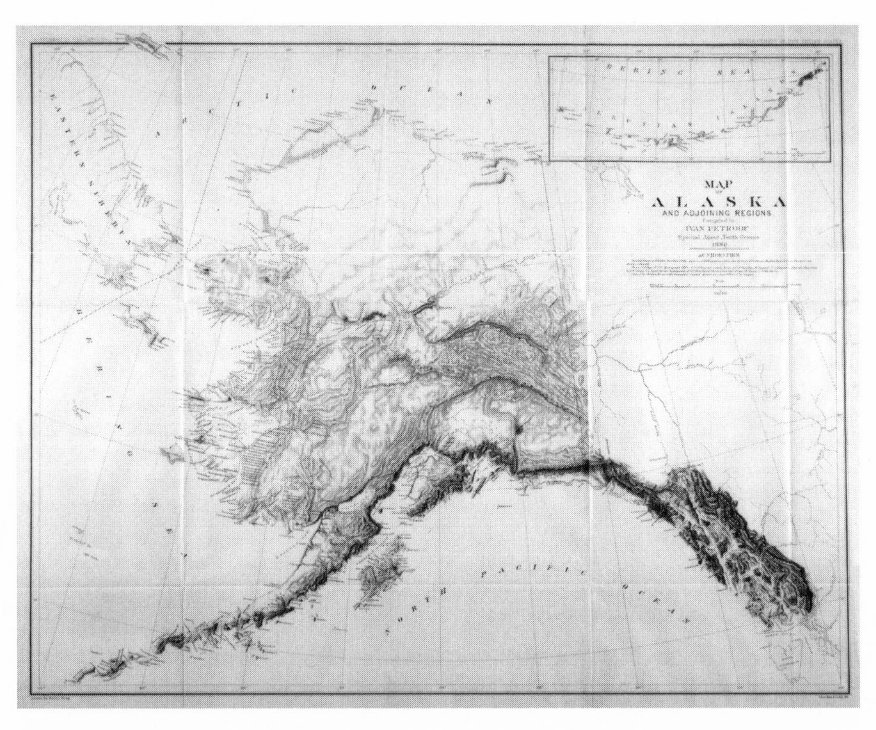

The maps published with the longer version of Petroff's 1880 census and *Report on Alaska* published in 1882 and reissued in 1884 included several maps that resembled the sketch map published in the 1881 summary report. All these maps appear to have utilized details of the upper Yukon and Tanana rivers, trails, villages, and posts in the region from the Kandik Map, though neither the map nor Kandik or Mercier are acknowledged in the credits. Ivan Petroff, *Report on Population, Industries and Resources of Alaska*, 1882.

On his return home Petroff was in demand for a few years, traveling to Washington to work on Bancroft's research and writing projects, and supplying his views on various northern issues to newspapers and academic gatherings. His paper on the interior geography and rivers of Alaska was read at the California Academy of Sciences meeting in March 1881 by the esteemed president of the group, George Davidson. The paper was entitled "Alaska's System of Inland Water Communication," and a newspaper article heralded Petroff's contributions: "Important Paper from Ivan Petroff on the Practicality of Reaching the Proposed United States Circumpolar Weather Station, at Point Barrow, by Canoes through Eastern Alaska."[425] According to Petroff, "Alaska bears the reputation of being an inaccessible country, as far as entry into its territorial limits is concerned. Thousands of miles of stormy seas must be traversed to reach its coast, but once the traveler sets foot on its soil he is greatly astonished at the great variety of avenues, in almost any given direction, that invite his progress. Roads do not exist, and even Indian trails are rare, but . . . [the] whole . . . interior is accessible by water communication . . . especially to natives with their canoes, or any traveler who will adapt himself to native customs."[426] The paper discussed the Kuskokwim, Tanana, and Yukon connections, stating white men did not travel these routes yet but Native people regularly used them to travel and trade.

Conclusion

The most striking feature of the Kandik Map, in view of all the interest in boundaries by the American traders and government officials, is the lack of a borderline. It is understandable in the context of the probable heritage of the mapmaker, "Yukon Indian" Paul Kandik. In 1880 he would have no reason to be concerned about the border, and no means of determining where it was located. Oral traditions,[427] historic records, and ethnographic and linguistic research[428] all document locations and land-use areas straddling the border for the different Athabaskan and Tlingit groups throughout this region.[429] These people today still consider that no boundaries exist between them of their own making. The 141st meridian was an arbitrary line drawn in 1825 with no reference to their traditional geography and groupings. It is notable that Mercier, with his interests in trade and nationality, while annotating the map with names of various trading posts along the river, did not attempt to draw a line on the map in the preferred or supposed location to support the needs of American trade companies, in which he was an active participant. Perhaps he simply had no time for such an exercise, nor scientific background and surveying skills to contribute to this issue. With no Canadian government officials on the scene to impose tariffs or other restrictions on the traders

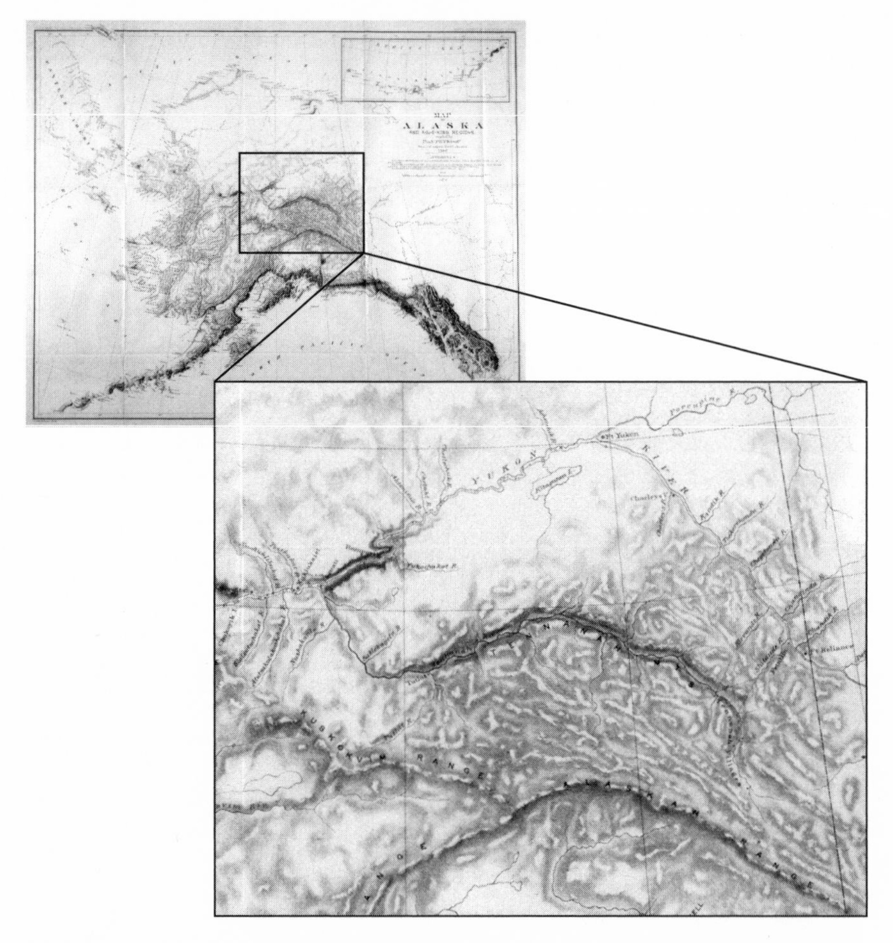

Detail of map published with Petroff's *Report on Alaska*, 1882 showing Yukon-Tanana confluence, Tanana villages, and trails between the rivers. The map is available as part of the Online Collections, "Meeting of Frontiers" rare maps Web site of the Alaska and Polar Regions Collections, Rasmuson Library, University of Alaska Fairbanks.

the boundary was still an irrelevant detail compared to the realities of daily life in the Alaska-Yukon borderlands.

While not all of Petroff's claims stood the test of time, his 1881 and 1882 *Alaska Census* reports and maps were cited for decades as later maps built upon the considerable data he collected. In particular the course of the Tanana and upper Kuskokwim, and connections between them and the Yukon, can clearly be seen as new additions to northern geographical knowledge that filled in gaps on Dall's maps and others. It is also very clear that this new information came from the Kandik Map with Petroff's cartographer copying the general shapes and relationships between the rivers and their tributaries from Kandik's drawing, along with Mercier's names. In future years, though, it is Petroff's name and not theirs that appears on the citations on subsequent maps such as the *U.S. Coast & Geodetic Survey Map of 1885*, Allen's map in 1886, and Canadian maps such as

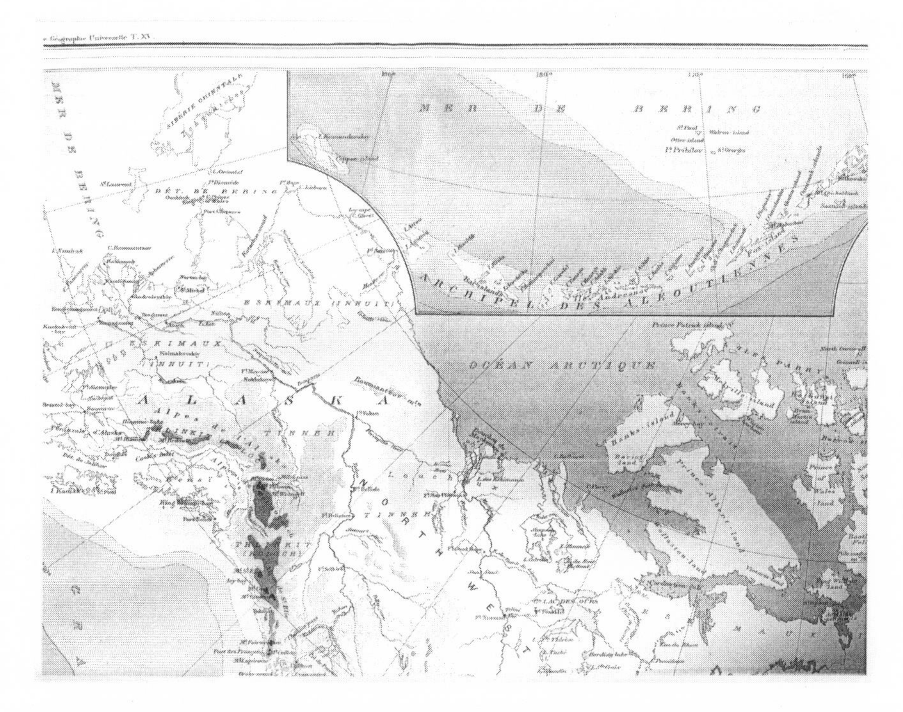

On this map of northwestern North America produced in France in 1890 François Mercier's influence is clearly recorded in the names of posts he established along the Yukon River—Fort Mercier, Fort Belle Isle, and Fort Reliance. Élisée Reclus, *Nouvelle Géographie Universelle La Terre et Les Hommes xv Amérique Boréale*, end map.

Johnson's compilation manuscript map of 1887,[430] which incorporated Ogilvie's surveys and cited Petroff among many sources. The British War Office cited Petroff on maps outlining various options for settling the vexing issue of the Alaska-Yukon boundary in the 1890s.[431] As further explorations occurred and more newcomers settled the area, additional details emerged and were recorded regarding the physical geography of the region, particularly with the work of Canadian surveyor William Ogilvie, and later American surveyors McGrath, Turner, and others.[432]

Mercier's exploits and francophone names appeared on French-language maps and in atlases such as Reclus' *Géographie Universelle*, published in Paris in 1890, which shows his namesake post Fort Mercier, as well as his other posts at Fort Reliance and Belle Isle. Gradually the hegemony of English place-names obscured the ancient Native ones, as well as Mercier's francophone renderings such as "R. Blanche" for the White River and "Chevreuil R." (Deer River in English) for the Klondike River.

On the eve of the Klondike Gold Rush interior Alaska and Yukon had been well traveled by Native people for countless generations, and the major river courses plus much else plotted by nonnative newcomers with their help for many decades. The discovery of gold along the Stewart and in the Fortymile creeks brought new urgency to document the travel routes from tidewater over the coastal mountains, and between the major watersheds, as well as the 141st boundary line, to complete the mapping of the Alaska-Yukon borderlands. The Kandik Map represents the high plateau of recorded knowledge about this region in 1880 and deserves this recognition. It is ironic that Paul Kandik's identity has been lost when so much was gained by so many others through the knowledge he shared so long ago.

Meetings and Meanings

My INITIAL contact with the Kandik Map dates back to 1983 when I was searching for the original Kohklux Map at the Bancroft Library in California, but the process of understanding the meanings of the map began when I arrived in the Yukon in 1974. I came to the Yukon to further my career as an archivist, having worked at the Ontario Archives where I developed a sense of archives primarily as "official" records, disconnected from any continuing community beyond the large nonpersonal aggregations identified as towns, counties, ministries, or the province as a whole. As a newcomer to the North, I needed maps, not only in a geographic sense but especially to understand my new community. Over the course of two decades, the Kandik and Kohklux maps have led me to people and communities very different from my own origins.

Two weeks after my plane landed in Whitehorse, I drove to Haines, Alaska, stopping at the Tlingit village of Klukwan and at Fort Seward to see the Chilkat Dancers perform. I was impressed by the pageantry of their regalia, the passion in their drumming, the intricacy of their dancing, and the sense of deeply rooted belonging presented in their clan stories and songs. Back at the Archives I looked for information about Tlingit people and was given a copy of Davidson's 1901 article about the Kohklux Map, showing Klukwan and the trail that became the highway I had just traveled. I retraced Kohklux's routes on current maps, increasing my curiosity about why he revealed his trails to the Yukon interior that was a fiercely defended Tlingit trading territory at the time. Given its size and significance, I thought Davidson must have preserved the original map somewhere. In 1983 I visited the Bancroft Library to search Davidson's papers. The library catalogues listed only his article, but when I explained my search to the map librarian he told me that more Davidson records had arrived in an old trunk recently, including two manuscript maps drawn by Kohklux.

The map librarian also told me of another northern Native map at the Bancroft Library and brought out the original Kandik Map. It had little documentation, being unconnected to any other collections or publications. It was a lone artifact in the huge Bancroft Map Collection, with no information on either Paul Kandik or François Mercier. To search for them I had to travel along many rivers and routes in the North, looking and listening for stories and meanings, paralleling their explorations and discoveries of long ago. In the process of finding my way through the landscape of their lives, my understanding of archives was transformed, from impersonal documentary sources to vital resources linked to community identities, past, present, and future.

Using Space and Knowing Place

Considering the backgrounds of the three known contributors, the Kandik Map has a remarkably complex set of influences embedded within it: from "Yukon Indian" Paul Kandik, who was born in the region and needed no maps to know where he was going, at least in his home territory; from French Canadian François Mercier, who was a southerner with more than a decade of travel and living experience in the region; and from Russian-born U.S. Census Agent Ivan Petroff, who made his first and only trip on the Yukon in 1880, probably with the map in hand to orient him as he traveled. In addition Petroff consulted with a number of the other traders such as McQuesten, Harper, and Mayo, and probably their wives. His Native paddlers would have added stories and details on the long trip upriver, and introduced him to other Native people along the way. All of these people certainly contributed to Petroff's knowledge and perhaps added details to the map itself. A number of the symbols on the Kandik Map and even some of the Native place-names included on it are puzzling to both Native and nonnative observers today, lost in the massive and pervasive changes that have swept through northern communities since the mapmakers' time. The transient nature of the nonnative people associated with the map has contributed further to the disassociation and dislocation of the map from its original communities of creation and knowledge, while providing opportunities for its preservation and dissemination. These ideas stimulate more questions that can be posed concerning the Kandik Map, ideas for contrast and comparison with other mapping traditions that introduce possibilities for understanding this map more fully.

How do people know a place, or find their way from one place to another and then know where they are in the new place? Tim Ingold, an English social anthropologist, analyzed these questions in his essay entitled "To Journey Along a Way of Life."[433] He contrasted the situation of a stranger in an unfamiliar area who needs a map to establish where he is or is going within a certain spatial or geographic location, and that of a person who grew up in a place who has no need of a map to get around because he moves through space based on previous travel experience. Some people theorize that the native person has a cognitive map "in his head," but Ingold discounts that idea and believes instead that history, experience, and movement provide the reference points for the native-born individual: "Places do not have locations but histories. Bound together by the itineraries of their inhabitants, places exist not in space but as nodes in a matrix of movement. I shall call this matrix a 'region.' It is the knowledge of the region, and with it the ability to situate one's current position within the historical context of journeys previously made—journeys to, from and around places—that distinguishes the countryman from the stranger. Ordinary wayfinding . . . more closely resembles storytelling than map-using. . . . [Map use or navigation plots] a course from one *location* to another in *space*. Wayfinding [is] moving from one *place* to another in a *region*."[434]

Ingold traces parallel activities for the countryman who knows a region and the stranger who uses a map, as both are moving along "paths of travel" and both of their journeys "unfold over time." However, the native's wayfinding should be distinguished from the stranger's navigation. The mapping of the former is also different from the mapmaking of the latter: "native maps . . . are not so much representations of space as condensed histories."[435] There are numerous theories to explain how people learn about wayfinding. Some analysts consider that the cognitive map derives from a prelearned understanding of features and locations and relations between them, so that travel from one point to another is simply a matter of matching movement to a mental pattern of places and spaces between them. Others, including Ingold, adopt the ideas of ecological psychology and consider that wayfinding is a "skilled performance" of a traveler feeling his way towards a goal through the ongoing monitoring of his surroundings.[436] Both approaches might result in some type of representation of place and space as a map.

People speak of maps in many different contexts, and Ingold considers a number of concepts to clarify what a map is and is not. He cites Alfred Gell's idea that "maps encode beliefs or propositions about the locations of places and objects that are true (or taken to be true) independently of where one is currently positioned in the world."[437] For example, the statement that London is south of Edinburgh is nonindexical; that is, its truth is not bound to the place where the statement is made. Gell defines a map as "any system of spatial knowledge and/or beliefs which takes the form of non-token-indexical statements about the spatial location of places and objects." However, a representation of the arrangement of peaks from a particular summit in a mountain range would be indexical of the place, an *image* by Gell's definition, because it is only valid from one point of reference and therefore indexical. Maps, especially modern topographical maps, purport to be nonindexical in Gell's sense. Ingold observes that "every map is embedded in a 'form of life,'" influenced by the "practices, interests and understandings of its makers and users."[438] Modern topographical maps require the viewer to be knowledgeable in the conventions, practices, and beliefs of the western science of cartography.

Ingold concludes that maps "index *movement*," and that "the vision they embody is not local but *regional*, but that the ambition of modern cartography has been to convert this regional vision into a *global* one, as though it issued from a point of view above and beyond the world."[439] A region in Ingold's view is every part of the world as it is "experienced by an inhabitant journeying from place to place along a way of life."[440] Ingold considers the paradox in his theory that people "know as they go," saying that "if knowledge of environment is embedded in locally situated practices, how is it constant when we move about? If all knowledge is context dependent how can we take knowledge from one context to another?"[441] People find their way en route and Ingold calls this process "mapping": "the traveler or storyteller who knows as he goes is neither making a map or using one. He is quite simply *mapping*."[442]

In many cultures stories about places serve important functions in addition to wayfinding or marking places for acquiring resources. Keith Basso studied Apache place-names and wrote that "American Indian place-names are intricate little creations . . . studying their internal structure, together with the functions they serve in spoken conversation, can lead the ethnographer to any number of useful discoveries . . . [and] a willingness to reject the widely accepted notion that place-names are nothing more than handy vehicles of reference. Place-names do refer . . . but in communities such as Cibescue, they are used and valued for other reasons as well."[443] Basso notes that Apache people, like their Navajo neighbors, travel frequently over great distances in regular patterns to and from their homes. They talk about place-names as a major part of their daily discourse and "habitually call on each other to describe their trips in detail."[444] With words they shape their massive physical landscape into a meaningful, remembered, and transmittable universe. Nick Thompson, one of Basso's Apache teachers, observed, "White men need paper maps . . . we have maps in our minds."[445] Basso relates a story about an Apache cowboy who uses place-names while repairing fences: "I like to [recite these names]. . . . I ride that way in my mind."[446] Apache names are semantically rich descriptive sentences for places of interest like "tseka' tu yahilii—water flows downward on top of a series of flat rocks."[447] Both Apache and Navaho are Athabascan languages, sharing many structural characteristics and even similar words in some cases with the Northern Athabaskan languages of Alaska and Yukon. The Hän name is similarly a "descriptive sentence" for Eagle Bluff. "Tthee t'äwadlenn" means "water hits the rock"—a very apt expression of this distinctive feature for orienting river travelers past and present. The name is documented both in Mercier's literate rendition of the name on the Kandik Map, as well as in Kandik's visual drawing of the feature at the confluence of the creek and the Yukon River.

Basso cites several other authors who have thought about similar connections between landscape, culture, and stories. Mikhail Bakhtin calls these locations "chronotopes—points where time and space intersect and fuse; a repository of distilled wisdom, a stern but benevolent keeper of tradition." Scott Momaday, a Native American author, also writes extensively about the importance of land to Native people as a "reciprocal appropriation" in which people feel deep connections with and responsibilities to their traditional places, and the importance of metaphor to understanding different cultural contexts like the relationship to landscape.[448] Basso emphasizes the importance of ethnography and linguistics for outsiders attempting to understand the metaphors of another culture: "To inhabit a language is to inhabit a living universe and vice versa."[449] These observations certainly apply to the linguistic elements of the Kandik Map which have provided insights into the origins of Paul Kandik and helped to identify the content and rationale of the map, while also presenting several

Tthee t'äwadlenn/Eagle Bluff is the most distinctive feature on this section of the river and a clear landmark for river travelers from ancient times to the present. It is shown on the Kandik Map as a distinctive hill shape. Photographer: Linda Johnson.

mysteries remaining to be resolved through ongoing discussions among Hän speakers, linguists, and other researchers.

Some people remain in their birthplace all their lives, but others are travelers and emigrants, venturing far from home to landscapes where they have no prior knowledge of indigenous languages, stories, or traditions to guide them. Their practices in wayfinding, marking, naming, mapping, and resettling offer more perspectives on the meanings of landscape. The means of travel, by foot or vehicle on land, in small or large craft on water, or in recent years, through the air by plane or spacecraft, affects the speed at which landscape is seen, the level of detail observed, and the opportunities for interpreting features and experiences en route. Each generation experiences landscape in the context of particular circumstances and therefore places evoke multiple layers of meanings over time. Kent Ryden begins his book *Mapping the Invisible Landscape* with a reflection on his experience of Route 101 at the border between Rhode Island and Connecticut. Over time earlier stone monuments, wooden signs, and metal markers have been lost in weeds at the side of the road, as the route was altered from footpath to horse-drawn carriage track to graded road for the first cars and now a multilane divided highway.[450] Message content and purpose have evolved as well in signage at this place, since travelers need more and different information now than the simple notification of the state boundary line offered long ago.

Ryden relates a different type of landscape experiences in the Coeur d'Alene mining district of Idaho. Newcomers from all over the United States and Europe traveled there over the past century to work in silver mines discovered in the 1880s. They established a new cultural identity that survives in stories passed on by the few permanent settlers, stories with personal connections or "memorates," rather than a history with a distant past.[451] The sense of local history is confined to what these newcomers "know as they go," in Ingold's terms. There is no reference to the linguistic origins of the district name, likely an artifact from the days of French Canadian fur traders like François Mercier, now abandoned in tourism literature in favor of "the Silver Valley," descriptive of the prevailing concept of the place, which is that "mining has always been the lifeblood of this valley."[452] Previous Native American occupants are only peripherally acknowledged in the "creation" story of the community today, which holds that the prospector deemed to be the "discoverer" of Coeur d'Alene was being "chased by Indians" when he found the silver ore body in 1883.[453] So people from all types of backgrounds and with varying amounts of time spent in a place use personal experiences to situate themselves. These stories become entrenched in the representations of community history over time, often replacing earlier memories, narratives, and names as happened with Kandik's and Mercier's linguistic traditions and identities on the Yukon River.

Ryden, Basso, and Ingold's conclusions that locally embedded knowledge about place derives from the history of movement and activity by people in a particular area fit well with the blended knowledge embodied in the Kandik Map. There appears to be a mixing of indigenous traditional knowledge and western cartographic information in both the details recorded and the sources for that knowledge. Paul Kandik functioned as a wayfinder in mapping his own homelands and as a cartographer interpreting others' knowledge for some sections. Some of the outer sections of the map are less detailed, probably because he was depicting these areas from the oral traditions of Tlingit or Tutchone people, not from "knowing the condensed history" through his own personal travels there. Some trails, such as Kohklux's coastal routes into the Yukon interior, may have been added by Petroff or others, including information from Tlingit people he met, or from George Davidson, or both. Mercier's role in attaching names to some of the places drawn by Kandik is revealing, too, as it appears he had only limited knowledge or memory of the tributary names in Hän country, and he added both French and English place-names for the upper Yukon that he had never seen, which originated with earlier Hudson's Bay Company traders. The outline of the Yukon, Pelly, and Porcupine river drainages on the Kandik Map resembles very closely the 1854 and 1857 maps produced by John Arrowsmith in London with information from Robert Campbell, as well as Dall's map published in 1875 with information from Raymond, Ketchum, and Laberge. The image of the Tanana, Kuskokwim, and mountainous areas in between them do not appear to be recorded on other published or manuscript maps that predate this map and likely were derived from Kandik's personal and community indigenous knowledge. So both the landscape images drawn by Paul Kandik and the names plus other "place" details recorded on the map by François Mercier, Petroff, and perhaps others were built upon the travel, experience, and stories shared among newcomers and indigenous residents, mixed with the interpretations and influences of cartographers in far-distant places. The Kandik Map is the result of a multigenerational, multicultural, and multipurpose gathering of knowledge about the landscape of the Alaska-Yukon interior region as it was known and understood by its creators in 1880.

Meetings, Meanings, and Motivations

While the motivations and needs of nonnative newcomers for geographic information seem obvious, why would Native people such as Paul Kandik share their knowledge with strangers? There are many examples of indigenous maps worldwide and a variety of styles, concepts, and methods for creating them. Among the examples of northern Native maps, a few have some related contextual documentation to clarify the approaches used by the mapmakers in visualizing and transferring their knowledge to a recorded format. This information has usually been assembled and interpreted by nonnative people so that it does not necessarily present a full understanding of the circumstances nor motivations and concepts for creating a particular map.

The closest example in time and place to the Kandik Map is the large Kohklux Map,[454] drawn in 1869 at Klukwan by the Tlingit chief and his wives. The story of this map was written by the American government surveyor George Davidson, who provided the materials for the mapping project, discussed the format, and wrote the Tlingit place-names on it, as directed by Kohklux and his two wives. Davidson wrote that the chief was inspired to demonstrate his geographic knowledge, and presented the map as a gift to him after the scientist predicted the exact timing of a solar eclipse that occurred in August 1869. The Chief and his wives spent three days drawing the outline of the rivers and mountains, discussing the naming and placement of tributaries, glaciers, and other features, all in Tlingit, which Davidson did not speak, so his interpretation of their conversations was limited. What is clear is that the Tlingit people recalled from memory a vast area of the interior Yukon through which they and their ancestors traveled repeatedly for countless generations to trade with Yukon Indian people. The format and north-south orientation of their map was probably influenced by George Davidson in several ways, because he supplied a coastal survey chart, the back of which served as the large piece of paper they requested for their drawing. Tlingit people previously drew sketches of their routes in the sand, but meeting George Davidson provided the first opportunity for Kohklux and his wives to use paper and pencil to represent their knowledge.[455]

The Kohklux Map story may have particularly close relevance to the production of the Kandik Map, since Petroff lived in San Francisco for several years, shared northern information and interests with Davidson on an ongoing basis, referenced Davidson's collection of data from a Tlingit chief, and very possibly saw Kohklux's maps before heading north to take the census in 1880. Davidson may have discussed the mapmaking process he witnessed and facilitated in Klukwan with Petroff as well. Davidson respected the Native people he met and their knowledge of landscape and travel routes,[456] which may have encouraged Petroff to seek out Native mapmakers on his northern travels. Since Petroff researched and copied a number of maps before he headed north and probably showed them to northerners he met, Kandik may have been inspired or encouraged to copy parts of them and fill in the blank areas from his own knowledge. Arthur Harper, Robert McDonald, and some riverboat captains had maps too so there could have been many occasions when the geography of the region was discussed and drawn out by Kandik and others during Petroff's visit.

Another more recent example of indigenous mapping in partnership with nonnative collaborators is Inupiat hunter Simon Paneak's drawings, texts, and maps, published with John Campbell as *North Alaska Chronicle: Notes from the End of Time, The Simon Paneak Drawings*. Paneak was several years old before seeing a *tunnik* or white person; his people, the Nunamiut, were interior tundra dwellers, decimated by white men's diseases in the late nineteenth and early twentieth centuries. In 1920 Paneak was one of a small group of survivors who immigrated to the north coast, then returned in the 1930s to their ancestral home in Anaktuvuk Pass. All the knowledge about places and resources had to be reintroduced to younger Nunamiut people, along with former religious, social, and political practices that sustained this lifestyle, as well as the ability to live and learn while traveling on the land. Campbell remembers first meeting Simon Paneak in a Nunamiut camp on the Arctic Divide in 1958, and from then until Paneak died at the age of seventy-five in 1975 "the two of us were friendly correspondents and fellow travelers." In the summer of 1967 Simon agreed to compose some drawings of Nunamiut history.

Campbell provided a sketchbook, pens, pencils, and crayons for the purpose but beyond that "Paneak was on his own. . . . Except for my requesting a tribal 'history,' a term that covers a lot of ground, no subject matter was suggested."[457]

What of Paneak's view of the work and the exchange? He clearly wanted to convey the beliefs and history of his people and drew on his store of mythological lore about the beginnings of time when giant animals flew through the air and mammoths roamed the tundra. Along with illustrations of tools and camp life, he made several maps of the North Slope based on Nunamiut subsistence patterns: "When . . . caribou failed to appear as predicted, at a particular time and place . . . the hunters, on the basis of their expert knowledge of the animal's habits, went looking for them in other quite specific places. Productive travel, both in good times and bad, demanded intimate knowledge of the terrain, with the result that a rather astonishing array of topographical features had names. . . . Men or family groups arriving in camp reported foraging and traveling circumstances according to precise geographical positions. . . . Though their maps were carried only in their heads, the Nunamiut not only knew their territory intimately, they had it mapped in detail as well."[458] Paneak's maps attest to his mastery of the landscape that was the source of his sustenance for a lifetime, and may have been drawn to establish a Nunamiut claim to the land as well. The white men with whom he worked used maps to record their data, so perhaps he wanted to demonstrate his ability to produce similar types of documents for what he knew so well. Paneak exemplified Ingold's theories about wayfinding being a "skilled performance," and his maps are indeed "condensed histories" of place, assembled over a lifetime of his own and his community's experiences in this "region."

By 1880 Paul Kandik would have witnessed the devastation of several epidemics along the Yukon River and the rapid changes overtaking the hunting, clothing, travel, and other aspects of life among his people. He may have been employed by traders and steamboat operators, gaining proficiency in one or more nonnative languages, and capable of assimilating many new opportunities, but like Paneak he may have been concerned about the future of his people and the loss of their cultural knowledge. Perhaps he had a similar pride and motivation to preserve his traditions and demonstrate his knowledge, and saw the power of paper and print as a communication tool through experience with traders, missionaries, and riverboat captains. Mercier and Petroff may have recognized him as a valuable adviser, as Paneak was by Campbell, someone who had traveled extensively, observed closely, and understood the needs and interests of his nonnative customers and employers.

There are many other examples of northern Native efforts to preserve and share information about landscape features and the unique knowledge associated with place-names and stories embedded in them. Katherine Peter's book *Neets'aii Gwindaii: Living in the Chandalar Country* provides an excellent example of how Native place-names help Gwich'in people to locate themselves in their traditional landscape and find the necessary resources for survival. Place-names refer to animal and plant food resources—like "Ch'ôonjik—Porcupine River," named for the abundance of that animal, which is easy to kill and serves as emergency survival food when all else fails; "Ch'at'oonjik—nest river," where eggs can be obtained; "Dyaach'i David Vattha'l—old man David's caribou fence," where caribou were hunted using corrals. Others point to firewood, medicinal resources, and trail markers: "Dinii Zhoo—brushy mountain"; "Tl'yahdik—a quarry for black rock used for medicine"; "Dachnlee—timberline" for hills southeast of Arctic Village; "Van Vats'an Hahdlaii—where a stream flows from the lake" near Arctic Village.[459] Peter wanted to ensure that this knowledge would be available to future generations, realizing that it is difficult now for young Gwich'in to learn their Native language and landscape.[460] In her attention to detailed descriptions of travel routes, resources, and camping places, Peter's work resembles Kandik's intensive representation of tributaries in Hän traditional areas, as well as his sketching of overland trails between watersheds.[461] While it is not possible to learn how Kandik acquired his landscape knowledge, his mapmaking efforts and purposes may in some respects parallel Katherine's collaboration with the Alaska Native Language Center. He used the unique opportunity of sharing his knowledge with Mercier and Petroff to document his region on a piece of paper, perhaps even with an idea that it would travel to faraway places and times via his nonnative colleagues.

Athabaskan people in the Fort Nelson region of northern British Columbia had exactly that idea in mind when they worked with anthropologist Hugh Brody during the 1970s to record their place names and stories associated with them in order to document their lifestyles and prepare testimonials for the Alaska Highway Pipeline Inquiry. In the face of impending economic developments that could potentially have devastated their traditional lands and hunter-gatherer seasonal rounds, Elders and other community members focused on the techniques used by outsider corporate researchers to map and present their proposals, then mapped and promoted their own vision of their lands and lives at the Inquiry hearings, in community meetings, and through mass media. In recent decades there have been numerous examples of such collaborative projects between indigenous people and nonnative researchers, providing all participants and those who experience the results of their projects with new tools to analyze space and place, and new opportunities to see mapping in different cultural contexts.

Barbara Belyea used Hugh Brody's description of Elder Joseph Patsah's approach to mapping as an example of the insights possible through intercultural exploration of these concepts: "The topographical map does not reveal the world to . . . Joseph Patsah; instead it is a prompt for the old man's memories and a link with continuing experience. . . . For Patsah, the area in question is a place where he has lived and hunted, rather than a space that is prescribed, circumscribed and possessed. He lays claim to it not by framing it within a boundary line on the map, but by a flutter of his hand in the air, a 'luminous track' that confirms his physical association with the land in the form of a gesture." She suggests that "we need to see Joseph Patsah's use of the topo sheets as a concession to scientific cartography—a complex response whose traditional and adaptive elements cannot ultimately be distinguished. . . . Traditionally, maps stored in memory, drawn from memory, had this guarantee of experience behind them. . . . The extent of such maps may well be greater than the country known personally to each cartographer. . . . The conformity of Amerindian maps also indicates a cultural memory at work. A Native map is never derived from personal experience alone, as our sketch maps are: the Missouri of the Blackfoot is a common image of that river, not invented on the spot by each mapmaker, but repeated from time to time and from cartographer to cartographer." Belyea asks us to consider what we can learn about earlier Native maps when we can't meet and converse with the mapmakers as Brody did with Joseph Patsah. She concludes, "Unfortunately missing the dialogue is missing the point. All we have left are the graphic transcripts, without explanations, memories, or associations. The maps are deprived of the 'luminous track'—the ephemeral structure of their significance."[462]

By comparison to Alaska and the Yukon, the landscape of the Yolngu people in Australia is vast and arid, but also focused on hunter-gatherer economics and technologies. Here too the difference between those who are native born and others who are newcomer travelers echoes with similar themes. The entire traditional territory in northern Australia, known to nonaboriginals as Arnhem Land, is imbued with profound meanings unique to Yolngu mythology and belief systems, captured in place-names in their own language and requiring an intimate knowledge of the stories associated with the names for interpretation. Several Yolngu bark paintings have been juxtaposed with Euro-Australian topographical maps to show both the relationship between the representations of physical landscape features and the contrasting worldviews of the respective cultures that created them. The Yolngu maps have vibrant colors, vivid delineations of rivers, bays, and juts of land, plus numerous mythological animals and humans, associated with symbolic representations of plants and other materials. They are recognizable as the same places depicted in Euro-Australian maps, but the uninitiated viewer is immediately aware that mysteries and meanings inhabit these maps far beyond the ken of nonaboriginals. There are no place-names in any form of literacy recognizable to print traditions. Yolngu people with the appropriate knowledge of clan symbols can read these maps as a history of place and people from creation through to present times. They are a perfect example of a unique and highly developed concept of landscape carried as "mental maps," passed on through many generations by oral traditions, sometimes represented in bark drawings in a form of documentary mapping, all of which are in large part closed to outsiders. Understanding these maps requires intensive knowledge of the physical landscape and the right to know the stories of the clans who create and own them.[463]

Paul Kandik does not appear to have imbued his map with any ritual signs or other cultural symbols, beyond the exclamation marks and other types of trail designations, the one "overturned canoe" symbol on the highlands between the Yukon and Tanana, and the boxes designating posts and camps, including those with flags. The flags may or may not have been his work—they are certainly emblematic of northern trading posts, appearing in drawings and photographs of all posts from H.B.C. through A.C.C. years. It appears that rivers, routes, and trading posts were the primary focus of this map, in keeping with the needs of its intended newcomer audience represented by Petroff. Like Kohklux, Kandik appears to have created a map in the style of the maps carried by the non-native traders, prospectors, and explorers he would have met and it seems likely he copied some parts of those maps too. Some of the Native places he included have well-known meanings and stories associated with them, such as Tthee t'äwdlenn (Eagle Bluff), while others have yet to be fully interpreted by Hän or other speakers. The map has served to preserve and bring forward these names, with the possibility of recovering more memories and stories in the future as Hän speakers and culture bearers continue the conversations necessary to learn about and from the map. The Kandik Map likewise serves as a documentary memory of the time when French Canadian conversations were a regular part of life along the Yukon River, together with Hän and other Athabaskan languages, as well as English and Russian.

While it is not possible to determine how Paul Kandik translated his "mental map" of his homeland region to paper, it is clear from Walter's map drawn in the snow for Dr. Kingsbury in 1890, through Katherine Peter's work, and many other examples, that northern Native people have a long tradition of "mapping" in Ingold's sense of recording movements and stories within their well-known and long-traveled region. As indigenous traditional knowledge gains more currency worldwide, there are increasing opportunities to explore the process of intercultural exchange, with more emphasis on Native speakers providing first-person narratives, rather than outside experts interpreting, editing, and shaping the picture of a culture that is not their own.

Stories of the Kandik Map

My journey with the Kandik Map has led to many people in many different places—each with a special perspective and set of knowledge to add to the story of what happened when Yukon Indian Paul Kandik collaborated with François Mercier from Québec to produce a map for American government official Ivan Petroff. Along the way I learned that an archival document was more than just a piece of paper with information. It is a repository of knowledge that comes alive in new ways with each person who sees it. I called upon all of my own background and innate understandings of homeland and frontier, of immigrant and native-born knowledge, of the fragility and persistence of heritage to travel along these paths.

Soon after I first saw the Kandik Map I met Anchorage researcher Linda Finn Yarborough, who was translating Mercier's manuscript memoirs at the time, and visited the Yukon looking for more information about him. I showed her my copy of the map and asked if Mercier had mentioned Paul in his writing. She had seen nothing of Kandik in her research, but she included the map in her publication as more evidence of Mercier's work in the North.[464] Attached to a modern publication, the Kandik Map started to receive more attention as the Mercier book circulated, though his relationship to Kandik was still a mystery.

The original Kandik Map journeyed back to the North for a Yukon Historical and Museums Association (Y.H.M.A.) conference in 1987, together with the two maps drawn by Kohklux. It received less attention than the other maps, perhaps because it was smaller and seemed less glamorous. It did intrigue several Hän people, young and old, whose traditional territories are the central focus of the map. Since the conference many people have puzzled over it, trying to identify Paul Kandik and determine the significance and meanings of his map.

Elder Percy Henry from Dawson has studied the place-names and tried to place the various tributaries shown on the Kandik Map.[465] Gerald Isaac, the grandson of Hän Chief Isaac, wrote a brief story about the first white people arriving at Hän villages and attached it to a copy of the map to stimulate discussion.[466] He gave a paper on the map at a conference in Dawson City in 1999, asking whether anyone had further information or ideas about its origins.[467] Tr'ondëk Hwëch'in researchers in Dawson City have conducted extensive oral history and archival projects in recent years, in conjunction with archaeology work at their traditional camps and efforts to reestablish links with their Hän relatives at Eagle and elsewhere in Alaska. The Kandik Map was discussed, but as with the Kohklux Map on the coast, no oral traditions appear to have been handed down concerning the creation of the map, nor any means of identifying the man Paul Kandik. Kingsbury's photos include several that document Hän communities in the 1880s, but today no one can verify the faces and relationships pictured in the photos.[468] The return of the documents in copy form to the North occurred just a little too late for identification by present generations.

When I entered the Northern Studies Masters Program at the University of Alaska Fairbanks in 2001 I decided to try and find more answers to the Kandik puzzles using the resources at the Rasmuson Library, together with the knowledge of Alaskan Native Elders and researchers whom I met while in Fairbanks. At the library, I saw an original copy of Ivan Petroff's 1882 *Census Report* with its maps of Alaska-Yukon river drainages that resembled Kandik's drawing. Reading his descriptions of maps and information obtained from traders and Indians, I began to think that Petroff was the instigator of the Kandik Map. I visited the Bancroft Library again in the summer of 2002 and read Petroff's correspondence and information on his employment by Bancroft. Then I looked at the original Kandik Map again and realized that the label that was previously stuck on the bottom was no longer attached to it, revealing the handwritten note stating that the map had been received at "St. Michael from M. Mercier." Comparing

Photo by RICHARD HARTMIER

CAREFUL HANDLING — The Klohklux map, drafted in 1869 and based on the memory of the coastal chief who invaded the Yukon to scare off fur trading competition from the Hudson Bay Company, went on display at the Yukon Archives today. It was brought here, from the Bancroft Library in Berkeley, and joins two other rare maps in a display to coincide with this weekend's annual convention of the Yukon Historical and Museums Association. Here archive worker Jim Connell wears gloves to handle the map, while an organizer of the display, historian Linda Johnson, looks on.

The Bancroft Library loaned the original Kohklux Map and the Kandik Map to the Yukon Historical and Museums Association for a conference in Whitehorse in 1987. Jim Lewis and Linda Johnson prepare the large Kohklux Map for display at the Yukon Archives. Photographer: Richard Hartmier.

the label and the note to Petroff's handwriting in other Bancroft sources proved them to be in the same script, confirming that he was directly connected to the map in some way, and suggesting that he had met at least one of its northern creators.

Documentary sources could illuminate a possible story line for the creation of the map and its subsequent provenance, but they held no further information about Paul Kandik's origins or his connections to François Mercier. I needed to learn more from people who knew the landscapes of the map and the life ways and history of the people who traveled through them in the past. I brought the map to Fairbanks Elder Effie Kokrine, born in 1920 on the Yukon near the mouth of the Tanana, and connected to the region where Kandik and Mercier might have met and certainly traveled. I showed her the photos of people and places that I had found at the Bancroft Library, which immediately resonated with her knowledge of Athabaskan lifeways, but she repeated what others had said about the faces being too distant in the past for people today to identify. In looking at the map, she noted that the greatest number of Native names were from upriver, not in her home territory at the mouth of the Tanana. She appreciated the grand sweep of territory covered by Kandik, noting that people could travel great distances because they were self-sufficient and knowledgeable about the land.[469] Later I met Dr. Phyllis Fast, a descendent of Seentahna (Jennie Bosco) and Arthur Harper, who was an associate professor at the University of Alaska Fairbanks.[470] She commented on Kandik's landscape imagery and also shared her family genealogy, adding to my sense of the significant role played by the Native wives of the early traders, and the difficulties in documenting their contributions.[471]

In Canada I met with Louise Profeit Leblanc, a Northern Tutchone woman originally from Mayo, who has a keen interest in the map. She had moved recently to Gatineau, Québec, and greeted me with the amazing news that her new neighbor was André Mercier, grandnephew of François and Moïse. She arranged for me to meet him and we reviewed the photos, map, and François' published *Recollections*. André said there were many sources on the Merciers in Québec and provided contacts for other family members. Many details of his uncles' fur trading years in the North had been handed down in the family, but he had not heard of Paul Kandik or seen the map.[472]

In Vancouver I visited another friend, Lulla John, whose late mother Nelna (Bessie) Johns was an Upper Tanana Elder with a vast knowledge of the borderlands depicted on the Kandik Map. Bessie was a speaker at several Yukon Historical and Museums Association conferences through the years and always stressed that Native people in Yukon and Alaska were "all one people," artificially divided by the imposition of the boundary at the 141st meridian.[473] I asked Lulla what the Kandik Map meant to her today as an Upper Tanana person. She said it made her proud of her First Nations ancestors, helping her to recall the many stories her mother told of traveling in the region, finding her way through stories passed on by her Elders about where to find berries, fish, caribou, and other resources, locating trails through particular creeks, lakes, and mountains or types of vegetation. Lulla hoped the map would be preserved and its story told so that future generations would be made strong and proud through knowing the skills of their ancestors.[474]

Back in Whitehorse I worked with linguists at the Yukon Native Language Centre to verify the Native place-names on the map, consulting

Linda Johnson traveled to Gatineau, Québec, in June 2006 to meet the great-nephew of François Mercier. André Mercier holds a copy of the Kandik Map that his great-uncle annotated, together with Louise Profeit Leblanc, a Northern Tutchone woman from the Yukon interested in the origins and stories related to the map. Photographer: Linda Johnson.

lists made decades earlier with people from Eagle and Dawson that clarified which rivers were shown and which were not, and the differences in linguistic forms used by Hän speakers from the two communities. I met with Gerald Isaac again to show him what I had learned so far about Petroff and Mercier, acknowledging that I had not been able to fill in many of the gaps that remained in identifying Kandik. He suggested that I travel to Eagle and Dawson to consult people there again since the evidence pointed more and more consistently to a Hän identity for Paul.[475]

I made the trip by road in August, and then by boat in September 2006, seeing both communities, as Paul Kandik must have done, as "part of the land" and "part of the water."[476] Being immersed in the landscape represented on the map was revealing in many ways, confirming that the map indeed has a rich array of detail, conveyed in subtle pencil shadings, which at first glance might be overlooked and underestimated. The trip on the river reinforced the skills needed to paddle traditional birch-bark canoes in the strong currents of the Yukon, and the knowledge required to identify creeks and rivers along the way. It was easy to see how essential Native guides were to early traders, how landmarks are plentiful and obvious, but needing distinctive names to be remembered, and how welcome steam travel must have been to cover the great distances and replace the huge energy required to paddle from place to place, especially upstream. At Dawson City and Eagle I asked various people about Paul Kandik, showing his map and the Bancroft photos, plus Petroff's maps. Researchers at the Tr'ondëk Hwëch'in Cultural Centre in Dawson City continue to display and study the map, but have not located any oral traditions or other means of identifying Paul Kandik. At Eagle I went to the Yukon-Charley Rivers National Preserve headquarters where I met Hän Elder Isaac Juneby, son of Willie Juneby, who supplied some of the place-name lists I used at the Yukon Native Language Centre. Isaac was working as an interpreter for the preserve and was familiar with the Kandik Map, which he had seen at the Upper Yukon River Heritage Symposium in Dawson City in 1999. He was quick to recognize the similarity between Kandik's and Petroff's maps, and immediately identified with the areas around Eagle and in the Kandik River valley, traveled by his family when his father was living. He quickly sketched out his own map of the area from memory, giving me an insight into how the same questions and responses might have transpired years ago between Kandik, Mercier, and Petroff.[477]

Eagle proved a fertile ground for discussing the Kandik Map. The Eagle Historical Society displays the map in their museum, sells Mercier's book, and maintains an extensive archives. There were no new sources on Paul, but I found a reference to Reclus' French geography text of 1890, which added significant new details to my file on François Mercier. At the old Eagle Indian Village I was invited to a community feast and again showed the map and photos, new to the people there and of great interest. Although they could not add to my information on Paul, they shared their knowledge of the landscape and names on the map.

On the way back to Dawson City on the Holland America catamaran *Yukon Queen II*, I met the ship's captain, Andy Bassich, who lived and trapped on the Fortymile uplands for years, and now lives in Eagle. Surrounded by the latest GPS and satellite imaging technology to navigate

Isaac Juneby is the son of Willie Juneby, who recorded Hän place-names in the 1970s that contributed to the identification of features on the Kandik Map. Isaac and his wife Sandy greeted visitors Queenie Copeland and Wanita Johnson at the Interpretive Center for the Yukon-Charley Rivers National Preserve in Eagle in 2006. Photographer: Linda Johnson.

Family, neighbors, and friends gathered at a feast at Eagle Village in September 2006. A copy of the Kandik Map and Dr. Kingsbury's photographs from 1889–1890 contributed to lively discussions of the history and people of the region. Photographer: Linda Johnson.

Captain Andy Bassich and Linda Johnson discuss the Kandik Map aboard the *Yukon Queen II* traveling on the Yukon River between Eagle and Dawson City in September 2006. Photographer: David Ashley.

the river, he was captivated by Kandik's drawing, appreciating its scope and detail, marveling at his ability to portray such a large area so well before the advent of sophisticated surveying technology. I asked him why he thought the Fortymile was left off the map and his reply was immediate—it is a dangerous river, with lots of rapids, difficult to navigate or hike alongside, while Mission Creek at Eagle is a much easier way to reach the upper Fortymile and cross over to the Tanana. He characterized the map as a chart "of routes and trails," not just rivers, which provided me with another way to think of the information Kandik presented as a system of travel options, rather than a simple rendering of geographic forms.[478]

In the fall of 2006 Y.H.M.A. sponsored a conference on northern research.[479] I presented my initial findings on the map, Kandik, Mercier, and Petroff, again pointing to the gaps and difficulties in locating information. This led to an invitation to meet with the Hän people at their Literacy Session at the Yukon Native Language Centre in Whitehorse in December 2006, where speakers from Dawson and Eagle were continuing work on recording place-names. With help from linguists they confirmed some of the names I had identified and added several more to their ongoing research projects. Hearing speakers from both Klondike and Eagle communities pronounce the place-names reinforced the idea that Mercier must have worked with Hän people from different areas during his years in the region, since his rendition of the names incorporates endings from at least two and possibly three language groups. The map provided a lively focus for remembering stories from the past, and perhaps one day such gatherings will lead to more information about Paul Kandik.[480]

The Y.H.M.A. conference also introduced me to some new sources on François Mercier as l'Association franco-yukonnaise had assembled newspaper articles, photos, and other sources from archives in Québec and in France on northern francophones. I met with historian Yann Herry and we compared notes about the Mercier brothers. These francophone connections led to more photos in Alaskan archival collections that filled in details of Mercier's life and provided identifications for some of his northern Native associates, which had not been available in English sources.[481] In Whitehorse, Parks Canada historian David Neufeld showed me copies of some H.B.C. maps and records, which further illustrated the longstanding tradition of information sharing between Native and nonnative travelers in the North.[482] Then I found some additional maps with Petroff attributions at Library & Archives, Canada, in Ottawa. A few weeks later I was back at U.A.F. and met Adeline Raboff, a Gwich'in speaker and author who has done extensive research linking oral traditions and documentary sources. She provided a thoughtful analysis of how and why the map might have been drawn, and explanations of Athabaskan naming traditions for places and people.[483]

In January 2008 I visited the U.S. National Archives II in Maryland to view the McGrath reports and found some more photographs taken by the American surveyors at the boundary in 1889–1891, but none as clear as those in the Davidson Collection albums and none of Native people in that area. The most exciting new development has been the appearance of Dr. Kingsbury's diaries with their stories of daily life at Camp Davidson, Forty Mile, and David's Camp. The identification and descriptions of "Big Paul" in the diaries and linkages to "Indian Paul, Pilot on the upper Yukon" in the photos taken by Kingsbury at the Indian camp

Research on the Kandik Map continues with Hän speakers from Dawson and Eagle Village exchanging ideas at the 2008 Moosehide Gathering. Left to right: Eagle Village resident Harry David, Jr., Elders Mabel and Percy Henry from Dawson City. Photographer: Linda Johnson.

near the surveyors' cabin lends more strength to the current conclusions about Paul Kandik's probable origins and life experiences as a Hän man, though still not offering certainty as to his identity or direct connections to descendents today.

Another recent "discovery" for me is the Alfred Boisseau's 1887 painting (see cover). In many ways it is an iconic expression of the differences between Kandik and Mercier—both past and present. Mercier was a brief traveler and resident in the region, yet he is commemorated in numerous sources today that confirm his identity and ensure his story will be circulated, valued, and remembered. His name is inscribed on the supply crate in the foreground of the painting—his certain claim to the North: "Francois Mercier, Alaska," evidence of his regard for that place and period, perhaps indicating that those years were the highlight of his life. He is depicted in the full regalia of a northern trader—fancy fur parka, snowshoes, whip, and with his sled and dogs prominently displayed on the river. Like Paul Kandik, the Native man in the painting is mysterious—unnamed, shrouded in darkness, his face obscured by the fur trim of his parka hood, seated behind and below Mercier, tending the tea kettle and fire—providing vital sustenance for his fellow traveler and himself. The backdrop to the painting resembles Tthee t'äwdlenn, recalling one of the photos taken by Charles Farciot of three Tanana men at Mission Creek in the early 1880s. No doubt Mercier provided the descriptions of the place, suggested the poses and positions of the people, animals, and equipment for the painting—his vision and version of his northern life and times.

Paul Kandik with his map, and Big Paul in his photos and the stories Kingsbury relayed about him, present an altogether different vision and version of these times. He represents the Hän people when they were still the majority culture on the upper Yukon, strong in their language and traditions, proud of their skills and knowledge, certain of their value and prepared to negotiate their own price for their services, capably self-sustaining within their kin groups, adapting to change, and contributing to the enterprises of their nonnative neighbors. He has left a proud legacy for his descendents today.

My journey with the *Kandik Map* is not complete—there are still more stories to hear and many documentary sources to check from Eagle, Fairbanks, and Dawson City in the North to Ottawa, Montreal, Washington, and San Francisco in the South—and who knows where else in the world there may be news of this map. By traveling through the same country as Kandik and Mercier, listening to the oral traditions of the past, gathering ideas about the map from people today, I have been able to recover some of its meanings. The challenge for me as a northern researcher, and for archivists and researchers everywhere working to preserve documentary heritage, is to stay long enough in the right places to hear and learn from the stories about the documents, past and present, and to preserve and present these multiple layers of context when interpreting the content and intent of the records.

Conclusion

Despite Petroff's meager estimate of the potential for gold discoveries, the 1880s along the Yukon evolved into a prospector's dream come true, with ever more exciting discoveries on the creeks between the Yukon and the Tanana. Ironically none of the early grubstakers from the fur trade era benefited from the ultimate Bonanza of 1896, and even the Native participants in the Klondike strike were Tagish people from the southern Yukon, not the home country of the Hän. After gold was discovered, the international boundary was of paramount importance and virtually all maps in subsequent years prominently featured the 141st meridian. In time it divided Native as well as nonnative residents with differing regimes for wildlife ownership and management, education, health care, and myriad other aspects of living in the modern world. The rivers with their migrating salmon, and the great herds of caribou traveling across the tundra, are among the few elements that still flow on without regard to the lines drawn in 1825, though they too are the subject of frequent international treaty negotiations, and sometimes disagreements, over appropriate management objectives and plans.

The Kandik Map remains as an enigmatic reflection on a unique period in time when Native and nonnative people traveled together down uncertain trails, unveiling the mysteries of northern geography for nonnative newcomers, and introducing new opportunities and challenges for indigenous residents. It is a visual metaphor for their differing perspectives on time and space in the interior of Alaska and the Yukon in the late 1800s. François Mercier spent seventeen busy years on the Yukon River.

In that short time he was an active agent of change, introducing significant technological, social, and economic innovations that had lasting effects long after he departed in 1885. His view of the region, as represented in his annotations on the map, focused on names and especially on the locations of trade posts along the river, which were his source of sustenance and learning on this new frontier. The posts are territorial representations, complete with flags to suggest a claim to some right of possession. Paul Kandik undoubtedly spent his entire life in the region. He was part of the place by birth, cultural heritage, and life experiences, adapting to new ways as opportunities came along, in the Athabaskan tradition. His part of the map is a continuous grand sweep of rivers flowing across Yukon and Alaska, without boundaries or flags of ownership. His sense of territory would have been derived from oral traditions evolving over many generations of travel, trade, and kinship, while his vision of the landscape he drew was probably a combination of remembering places from numerous repetitive trips paddling on the water, hunting and gathering on the land, and traveling to trade rendezvous in various locations. The extensive area he represented is evidence of the fluidity of geographic knowledge across northern Native trade and travel networks. It also suggests the introduction of steamboats may have offered Paul Kandik a broader range of travel in Yukon and Alaska than was possible for his ancestors.

While the Kandik Map today holds few geographical surprises, it has great potential as an enduring messenger among people in the North, bringing forward knowledge of landscape, languages, and ways of learning from the past to bring together and inspire present and future generations. Despite all their differences, Kandik and Mercier shared common places, common journeys, and perhaps even common kin. Their cooperative efforts in drawing the map without a boundary at the 141st beckon to people today to consider their common interests, obligations, and future together in a landscape connected by great rivers of change.

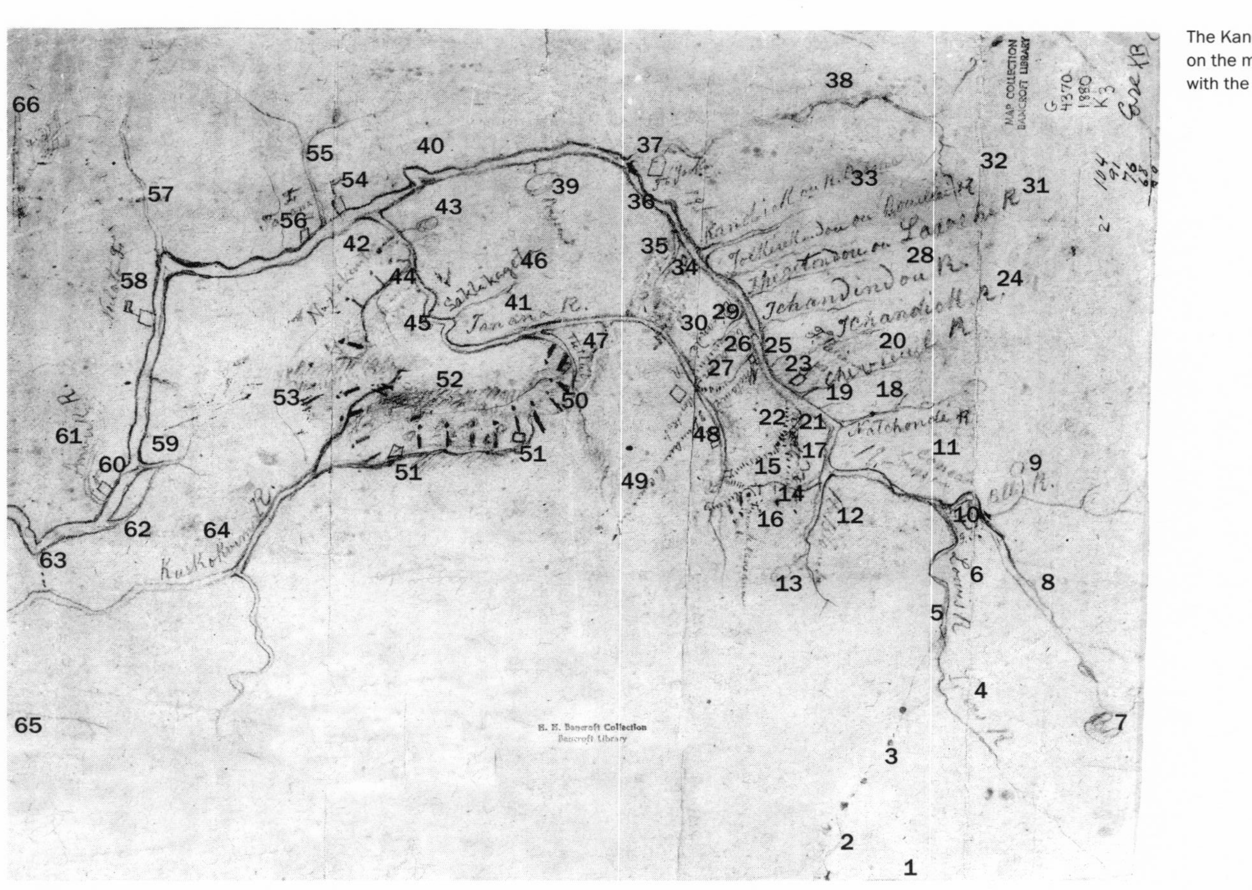

The Kandik Map. Numbers on the map correspond with the following chart.

Appendix: Chart of Place-Names

	Place-Names as Recorded on Kandik Map (most by Mercier)	*Hän or other Athabaskan Language Place-Name*	*Current Official or Common Usage Toponym*	*Comments*
1	No name		Lynn Canal	
2	Chilkat R.		Chilkat River	
3	No name		Kohklux or Dalton Trail	This series of dots and dashes denotes a trail from Chilkat River to Lake Laberge (?) similar to Tlingit trade routes to interior Yukon depicted on the large Kohklux Map.
4	Lewes R.		Yukon River	Lewes River was the name applied to the upper Yukon above Fort Selkirk by early Hudson's Bay Company traders and other explorers.
5	No name	Ta'an Mün	Lake Laberge	W.H. Dall named this lake on his 1875 map after his French Canadian colleague and fellow explorer Michel Laberge, but the name was not in common usage in the North in 1880 when the Kandik Map was drawn.
6	Louis R.		Yukon River	Mercier included a French version of the name "Lewes River" on the section of the river above Fort Selkirk.
7	No Name		Frances Lake	The distinctive "C" shape representing this large lake on the upper Pelly River in southeast Yukon appeared as a similar rendition on many early maps of the region.
8	No name		Pelly River	Mercier attached the Pelly River name to the McMillan tributary instead of what is now named as the main river.
9	Pelli R.		McMillan River (?)	Mercier spelled the name of this river with an "i" instead of its English version with a "y," as designated by Robert Campbell in honor of a Hudson's Bay Company official.
10	Ft. S.		Ft. Selkirk	A square box with a flag identifies this Hudson's Bay Company post first established by Robert Campbell in 1848 at the confluence of the Pelly and Yukon rivers, shown on this map in its first location on the south bank of the Pelly River.

Appendix: Chart of Place-Names, continued

	Place-Names as Recorded on Kandik Map (most by Mercier)	*Hän or other Athabaskan Language Place-Name*	*Current Official or Common Usage Toponym*	*Comments*
11	11 caneau jours			This annotation appears to denote traveling days by canoe on the river, perhaps from Fort Reliance upriver to Fort Selkirk.
12	R. Blanche		White River	
13	No Name		Wellesley Lake (?)	
14	No name	Khel ndek (?)	Sixtymile River (?) or Ladue River (?)	Khel ndek is the Hän name for the Sixtymile River but it is not clear which river is depicted on this map.
15	No name		Hän-Tanana or Sixtymile River Trails	This series of symbols resembling exclamation marks denotes the trails between the upper Ladue or Sixtymile tributaries of the White River and the upper Tanana River, identified in Hän oral traditions as traditional routes for travel and trade.
16	No name			This symbol appearing as a dot with a half circle over it is unknown in meaning along with the indecipherable words beside it.
17	No name			This symbol resembling an inverted "M" is of unknown meaning.
18	Natchondé R.	Nächòo ndek	Stewart River	Kandik drew two lakes on the upper Stewart River which are not named.
19	No name	Dhänt'all ndek (?)	Indian River (?)	
20	Chevreuil R.	Tr'ondëk or Tr'odëk (Klondike Hän speaker) or Tr'oju (Eagle Hän speaker)	Klondike River	*Chevreuil* means "deer" in French and other early maps of the region designated this as the Deer River.

Appendix: Chart of Place-Names, continued

	Place-Names as Recorded on Kandik Map (most by Mercier)	Hän or other Athabaskan Language Place-Name	Current Official or Common Usage Toponym	Comments
21	Katcé (?)		Nu-Kla-to (various spellings) and/or Catsah's Camp	This very faint name on the tributary across from Fort Reliance was the location of a traditional camp of "Katcé" as Mercier spelled his name, the Hän leader who piloted him upriver in 1874. He was known as "Chief Catsah" (with various spellings) to Campbell, McQuesten, and other early explorers.
22	No Name		Hän-Tanana Trails	This series of hatch marks denotes trails between Catsah's Camp on the Yukon and the upper Tanana River and the Sixtymile or Ladue rivers, identified in Hän oral traditions as traditional routes for travel and trade.
23	F. Raliance		Ft. Reliance	A square box with a flag identifies this post established by François Mercier in 1874 across from Catsah's Camp with the help of Hän people and left in charge of traders McQuesten and Banfield.
24	Tchandick R.	Tthee t'ak ndek	Chandindu or Twelve Mile River (?)	Mercier appears to have recorded the name for this river with a "dick" ending similar to the "dek" ending used to denote a water course by Klondike River Hän people, while the name used today resembles the "juu" or "dou" ending as used by Eagle Hän speakers.
25	No name		Belle Isle (?)	The square box drawn at this site could denote "David's Camp" named after the Hän leader known to the white traders as "Chief David," or Mercier's "Belle Isle" trading post which was located close to the camp.
26	Thetawdé	Tthee t'äwdlenn juu	Mission Creek	The Hän name for the prominent bluff at the mouth of this creek is "Tthee t'äwdlenn" meaning "water hits the rock" so Mercier appears to have applied his version of the Hän name to the creek. The bluff is shown by Kandik as a distinctive inverted "V" at this point on the river.
27	No name		Hän-Tanana Trails	This series of hatch marks denoting trails between David's Camp and the villages of Upper Tanana people on the Tanana River are identified in Hän oral traditions as traditional routes for travel and trade.

Appendix: Chart of Place-Names, continued

	Place-Names as Recorded on Kandik Map (most by Mercier)	Hän or other Athabaskan Language Place-Name	Current Official or Common Usage Toponym	Comments
28	Tchandindou R.		Eagle Creek (?)	This name recorded by Mercier is puzzling since the tributary bearing a similar name today is upstream on the east bank, the one he labeled as “Tchandick R.”
29	Klevandé R.	Tl’evär juu	Seventymile River	Mercier recorded his version of a Hän name for this river. “Tl’evär tthèe’” is the Eagle Hän name for “Calico Bluff,” which is a prominent landmark near the mouth of the Seventymile River.
30	No name		Hän-Tanana Trails	This series of hatch marks between the Seventymile River and the upper Tanana River are identified in Hän oral traditions as traditional routes for travel and trade.
31	Thegetondou ou Laroche R.	Thee taw juu (?)	Tatonduc River (?)	Mercier recorded his rendition of a Hän name for this river and an alternate French name prefaced with “ou,” which translates as “or” in English, meaning white traders may have used both names. Translated to English “Laroche” means “Rock River” and may have referenced a landscape feature or perhaps a person.
32	Tolkesekandou ou Bouillie R.	Ddhäw khèww juu (?)	Nation River (?)	Mercier recorded his rendition of a Hän name and an alternate in French prefaced with “ou,” which translates as “or” in English, meaning white traders may have used both names for it. Translated to English “Bouillie” means “swirling” and may have referenced a landscape feature or perhaps a person.
33	Kandick ou R. Calme	K’ày’juu (Eagle Hän speaker)	Kandik River or Charley Creek	Mercier recorded his rendition of a Hän name, which means “Willow Creek,” and an alternate in French prefaced with “ou,” which translates as “or” in English, meaning white traders may have used both names for it. Translated to English “R. Calme” means “Calm River” and may have described the nature of the river.
34	No name		Chief Charley’s Camp	These two small square boxes on the west bank of the Yukon River opposite the Kandik River probably denote camps used by Chief Charley’s Band in the 1870s and 1880s.
35	Erased name	Tr’e’òk juu; Tr’äkhö’oll juu (?)	Charley River (?)	The name recorded on the map has been erased but this is almost certainly the Charley River.

Appendix: Chart of Place-Names, continued

	Place-Names as Recorded on Kandik Map (most by Mercier)	*Hän or other Athabaskan Language Place-Name*	*Current Official or Common Usage Toponym*	*Comments*
36	No name	Tth'echù; Chuu k'onn (Hän names for Yukon River)	Yukon River	Mercier did not record the name "Yukon River" anywhere along its course but he certainly would have heard the Hän name for the upper river spoken in various ways by the people at Charley's, David's, and Catsah's camps.
37	F. Yukon		Ft. Yukon	A square box with a flag identifies this Hudson's Bay Company post first established by Alexander Hunter Murray in 1847, later occupied by American traders starting with Moïse Mercier in 1869, and by Gwich'in Chief Shahnyaati' in the mid-1880s.
38	No name		Porcupine River	The Porcupine River is shown with only one tributary, which may be the Bell River, and some dots to the south of the Porcupine, which may indicate some Gwich'in camps.
39	Poissons		Fish Lake (?), may be Birch Lake (?)	Kandik drew a large lake downriver from Fort Yukon and Mercier labeled it "Poissons," which translates as "fish" in English. This could be one of many lakes used by Gwich'in people as fisheries to supply their own needs and for trade to the Hudson's Bay Company and later traders, and perhaps is the Birch Lake of today.
40	No name		Chief Shahnyaati' Camp	This prominent bend in the Yukon River was identified on many early maps as one of the principal camps of the renowned Gwich'in Chief Shahnyaati', whose name was spelled "Senaté" by Mercier and various ways by other traders.
41	Tanana R.		Tanana River	The Kandik Map is the earliest map found to date which shows the upper course of the Tanana River, including villages of the Upper Tanana people who traveled downriver to trade at Nuklukayet and Fort Yukon and on overland trails to the Yukon to trade at Belle Isle and Fort Reliance.

Appendix: Chart of Place-Names, continued

	Place-Names as Recorded on Kandik Map (most by Mercier)	Hän or other Athabaskan Language Place-Name	Current Official or Common Usage Toponym	Comments
42	Nuyklukaut		Nuklukayet (?)	The lettering of this name appears to be Petroff's rather than Mercier's and probably refers to the traditional trading area at the Tanana-Yukon confluence where Gwich'in, Tanana, and other groups gathered to exchange goods before and after the arrival of white traders.
43	No name		Minto Lakes	This large lake was not named but is almost certainly Minto Lakes of today, the location of another well-known fishery shown and named on Raymond's and Dall's maps.
44	No name		Kantishna River (?)	The one unnamed tributary of this river may be the Toclat River shown on other early maps of the area.
45	No name		Tanana camp (?)	This square box may denote a Tanana camp or possibly the location of Mr. and Mrs. Bean's home where she was murdered in 1878. It is close to the location identified by McQuesten as "Short Station" and by Lieutenant Allen as "Harper's Station" and may be intended as that post, though located on the wrong side of the Tanana.
46	Saklikegata		Salcha River (?)	The lettering of this name appears to be in Petroff's hand rather than Mercier's.
47	Tutluk		Delta River (?) or Tok River (?)	The lettering of this name appears to be in Petroff's hand rather than Mercier's.
48	No names		Upper Tanana villages	A series of five square boxes on the upper Tanana River likely represent the camps of Upper Tanana people.
49	3 jours avec le caneau		Canoe Route to Yukon River (?)	This annotation is written in Mercier's hand beside the Tanana River at the western end of one of the Hän-Tanana trails marked in the uplands of the Fortymile River. It may denote the travel time to the Yukon from this point or between villages on the Tanana River.

Appendix: Chart of Place-Names, continued

	Place-Names as Recorded on Kandik Map (most by Mercier)	*Hän or other Athabaskan Language Place-Name*	*Current Official or Common Usage Toponym*	*Comments*
50	No name		Upper Tanana Trail (?)	A series of inverted exclamation marks appears to denote a trail connecting the upper and lower Tanana River with the upper branches of the Kuskokwim River.
51	No names		Upper Kuskokwim camps (?)	Two small square boxes on the southern branch of the upper Kuskokwim River may denote camps of Upper Tanana people or other Native groups, or of white traders.
52	No names		Alaska Range (?)	Extensive shadings in the area to the west of the Tanana River probably represent the Alaska Range of mountains. The Upper Tanana people were known as the "men of the mountains" or "gens de butte" in the early years of Hudson's Bay Company trade.
53	No name		Denali (?); Mount McKinley (?)	A series of three prominent lines may represent the mountain known as Denali or Mount McKinley, well known to all travelers in this region, past and present.
54	No name		Fort Mercier (?)	This square box with a flag may represent the post Mercier built in 1877 for the Western Fur and Trading Company and named after himself. He dismantled the building in 1884 and and rebuilt it downriver near the other Alaska Commercial Company buildings at the Nuklukayet post.
55	No name		Tozitna River (?)	
56	F. Tanana		Fort Tanana; Tanana Station	This square box with a flag could be one of several posts established in this vicinity after 1869 when Mercier established Fort Adams, later referred to as Nuklukayet, and operated by a number of different traders for various trading companies through the years. In 1880, when the Kandik Map was drawn, it was operated by Harper and Mayo and their families for the Alaska Commercial Company.
57	No name		Koyukon River	

Appendix: Chart of Place-Names, continued

	Place-Names as Recorded on Kandik Map (most by Mercier)	*Hän or other Athabaskan Language Place-Name*	*Current Official or Common Usage Toponym*	*Comments*
58	Nulato F.		Fort Nulato	A square box with a flag identifies this post operated by different traders for various companies through the years. Mercier may have resided here at some point and may have had a child with a woman at this community.
59	No name		?	Unnamed tributary of lower Yukon River.
60	No name		Fort Anvik; Anvik Station	A square box with a flag identifies this post operated by different traders for various companies through the years.
61	Anvick R.		Anvik River	
62	No name		?	Unnamed tributary of lower Yukon River.
63	No name		Trail or portage (?)	This series of dots depicts a trail or portage between the lower Yukon and the Kuskokwim river.
64	Kuskokwim R.		Kuskokwim River	The main course of the lower Kuskokwim River is shown with one unnamed tributary flowing to the south.
65	Rec'd from M. Mercier at St. Michael, July 1, 1880			This annotation on the Kandik Map is only visible when the paper label identifying Paul Kandik and François Mercier as the creators of the map that was taped to the original map is removed.
66	No name		Saint Michael (?)	This very faint depiction of a large fort with two large boats may represent Saint Michael on the Bering Sea.

Native place names from Willie Juneby, Unpublished manuscript, "Place Names of the Eagle Region" transcribed by John Ritter August 1978, typescript at Yukon Native Language Centre, Whitehorse, and Jules Jette, *Koyukon Athabaskan Dictionary*.

Notes

1. Paul Kandik and François Mercier. "Map of Upper Yukon, Tananah and Kuskokwim rivers 1880." Cited hereafter as the "Kandik Map." Map Collection, G4370 1880 K3 Case XB. The Bancroft Library, University of California, Berkeley.

2. Allen A. Wright, *Prelude to Bonanza* (Whitehorse, YT: Studio North, 1992), 1–117.

3. James Wickersham, *Old Yukon Tales—Trails—and Trials* (Washington, DC: Washington Law Book Co., 1938), v, 446–7.

4. Hubert H. Bancroft, *History of Alaska 1730–1885* (New York: Antiquarian Press Ltd., 1959), xxiii–xxxviii.

5. Ken S. Coates and William R. Morrison, *Land of the Midnight Sun: A History of the Yukon* (Edmonton, AB: Hurtig Publishers, 1988), 320–27.

6. Linda E. T. MacDonald and Lynette R. Bleiler, *Gold & Galena: A History of the Mayo District* (Mayo, YT: Mayo Historical Society, 1990), xi–xvi, 3–31.

7. Julie Cruikshank and Jim Robb, *Their Own Yukon: A Photographic History by Yukon Indian People* (Whitehorse, YT: Yukon Indian Cultural Education Society and Yukon Native Brotherhood, 1975), v–vi, 13–179.

8. Carol Geddes, *Picturing a People: George Johnston, Tlingit Photographer*, videocassette (Montreal, QC: National Film Board of Canada in coproduction with Nutaaq Média Inc. and Fox Point Productions, 1997).

9. Yukon Native Brotherhood, *Together Today for Our Children Tomorrow* (Whitehorse, YT: Yukon Native Brotherhood, 1973).

10. Angela Sidney, Kitty Smith, and Rachel Dawson, *My Stories Are My Wealth* (Whitehorse, YT: Council for Yukon Indians, 1977), I–III.

11. Adeline Peter Raboff, *Inuksuk Northern Koyukon, Gwich'in & Lower Tanana 1800–1901* (Fairbanks, AK: Alaska Native Knowledge Network, 2001), 3–46.

12. Thomas R. Berger, *Northern Frontier, Northern Homeland: The Report of the Mackenzie Valley Pipeline Inquiry*, Vol. 1 (Ottawa, ON: Canada, 1977), vii–xxvii.

13. Thomas R. Berger, *Village Journey: The Report of the Alaska Native Review Commission* (New York: Hill and Wang, 1985), 5–19.

14. Julie Cruikshank, *Life Lived Like a Story* (Vancouver: University of British Columbia Press, 1990), 1–20.

15. Cécile Girard and Renée Laroche. *Un Jardin sur le Toit: La Petit Histoire des Francophones du Yukon* (Whitehorse, YT: L'Association franco-yukonnaise, 1991).

16. Judy Ferguson, *Parallel Destinies: An Alaskan Odyssey* (Big Delta, AK: Glas Publishing Company, 2002), 7–30.

17. Sha Tan Tours, *Yukon First Nations Map: The Real Map According to Our Elders* (Whitehorse, YT: Sha Tan Tours, c. 1998).

18. Julie Cruikshank, *The Social Life of Stories: Narrative and Knowledge in the Yukon Territory* (Vancouver: University of British Columbia Press, 1998), 71–81.

19. William Ogilvie, *Early Days on the Yukon* (Ottawa, ON: Thornburn & Abbott, 1913), 119–36.

20. Pierre Berton, *Klondike, The Last Great Gold Rush* (Toronto, ON: McClelland & Stewart Limited, 1972), 37–47.

21. Catharine McClellan, "Wealth Woman and Frogs among the Tagish Indians," *Anthropos* 58 (1963): 121–8.

22. Coates and Morrison, *Land of the Midnight* Sun, 79–82.

23. H. A. Cody, *An Apostle of the North: Memoirs of the Right Reverend William Carpenter Bompas, D.D.* (Toronto, ON: The Musson Book Co. Limited, 1908), 53–142.

24. Frederick Schwatka, "Report of First Lieut. Frederick Schwatka, 'Reconnaissance of the Yukon River, 1883,'" and Henry Allen, "Report of Lieut. Henry Allen, Second Calvary, U.S.A. 'A military reconnaissance of the Copper River Valley, 1885,'" in *Compilation of Narratives of Exploration in Alaska*, Senate Committee on Military Affairs, Senate Report 1023, 56th Congress, 1st session, 1900 (Washington, DC: U.S. Government Printing Office, 1900), 285–362, 411–88.

25. Wright, *Prelude to a Bonanza*, 118–301.

26. Morgan Sherwood, *Exploration of Alaska 1865–1900* (Fairbanks: University of Alaska Press, 1992), 106–18.

27. Ibid., 115.

28. Ibid., 118.

29. Allen, "Report of Lieut. Henry Allen," 409–94.

30. Wickersham, *Old Yukon Tales*, 313–5.

31. David Neufeld, "The Commemoration of Northern Aboriginal Peoples by the Canadian Government," *The George Wright Forum* 19, no. 3 (2002): 22–33.

32. Katie John, "Nen'k'e Tezyaade, When Lieutenant Allen Came into the Country," in *Tatl'ahwt'aenn Nenn' The Headwaters People's Country: Narratives of the Upper Ahtna Athabaskans* told by Katie John . . . [et al.]; transcribed and edited by James Kari; translated by Katie John and James Kari (Fairbanks: Alaska Native Language Center, 1986), 115–86.

33. Craig Mishler and William E. Simeone, *Hän Hwëch'in People of the River* (Fairbanks: University of Alaska Press, 2004), xvii–xxxii.

34. The names of both Paul Kandik and François Mercier appeared in Google searches in February 2007 but Kandik's name is only associated with the map, whereas

biographical information appears for Mercier. In addition, online searches for Mercier produced photographs and other details through the Web site of the Bibliothèque et Archives nationales du Québec.

35. Kent Ryden, *Mapping the Invisible Landscape: Folklore, Writing, and the Sense of Place* (Iowa City: University of Iowa Press, 1993), 174.

36. Google references to the Kandik Map include a paper by Catharine McClellan given at the 1987 Yukon Historical and Museums Association conference in Whitehorse entitled "Before Boundaries, People of Yukon/Alaska," located at http://www.yukonalaska.com/yhma/public/mcclell.html, and two online Yukon Archives displays, "At Home in the Yukon" and "Yukon with a French Touch," located at http://www.tc.gov.yk.ca/archives/frenchyukon/voyag/voyag2.html, plus several other citations.

37. Sherwood, *Exploration*, 57–58.

38. Various stamps identify the map as part of The Bancroft Map Collection, and handwritten annotations provide reference coordinates for its location in the collection.

39. Cornelius Osgood, *The Han Indians: A Compilation of Ethnographic and Historical Data on the Alaska-Yukon Boundary Area* (New Haven, CT: Yale University Press, 1971), 20.

40. Marvin Falk, *Alaska World Bibliographic Series*, vol. 183 (Oxford: Clio Press, 1995), 4–56.

41. Marvin Falk, "European Images of Pre-discovery Alaska," in *Unveiling the Arctic*, ed. Louis Rey, 562–73 (Calgary, AB: The Arctic Institute of North America, 1984).

42. Catharine McClellan, *Part of the Land Part of the Water: A History of the Yukon Indians* (Vancouver, BC: Douglas & McIntyre, 1987), 3–15, 105–7, 233–9.

43. Wright, *Prelude to a Bonanza*, 1–117.

44. Ibid.

45. Lewis Green, *The Boundary Hunters Surveying the 141st Meridian and the Alaska Panhandle* (Vancouver: University of British Columbia Press, 1982), 1–6.

46. Wright, *Prelude to a Bonanza*, 280.

47. François Xavier Mercier, *Recollections of the Youkon: Memoires from the Years 1868–1885*, ed. Linda Finn Yarborough (Anchorage: The Alaska Historical Society, 1986).

48. Helen Dobrowolsky, *Hammerstones: A History of the Tr'ondëk Hwëch'in* (Dawson City, YT: Tr'ondëk Hwëch'in, 2003), 3–18.

49. The title on the document is "Map of the Upper Yukon, Tananah and Kuskokwim rivers"; however, it is known more commonly in the Yukon today as "The Kandik Map."

50. Ibid., 118–28.

51. David Kessler, reference librarian at the Bancroft Library, conversation with the author, Berkeley, California, April 2006. Kessler confirmed that no information is available concerning the acquisition of the map by the library.

52. Thomas J. Turck and Diane L. Turck, "Trading Posts along the Yukon River: Noochuloghoyet Trading Post in Historical Context," *Arctic* 45, no. 1 (March 1992): 51–61.

53. Gerald Isaac, "The Han Huch'inn Early Warning System," June 9, 1989 (ms. in personal collection of Linda Johnson).

54. National Park Service, *Yukon-Charley Rivers National Preserve* (Washington, DC: U.S. Government Printing Office, 2005), map on back of pamphlet.

55. J. Arrowsmith, "British North America 1854 Reproduction" (Ottawa, ON and Whitehorse, YT: Association of Canadian Map Libraries and Yukon Historical and Museums Association, 1982). Dall also recorded the name "Antoine River" in a similar location on his 1875 map of Alaska and Yukon. The name Antoine was probably a reference to Antoine Houle, an H.B.C. employee and interpreter at Fort Yukon for many years.

56. Mercier, *Recollections*.

57. Ivan Petroff to his wife, May 10, 1881, Alexandrovski, Cook's Inlet, Alaska. Petroff Correspondence, The Bancroft Library, University of California, Berkeley. Comparison of the letters "I," "K," "M," and "P" in the letter with the annotation on the lower left corner of the map, and the annotation on a separate piece of paper attached to the map until recently, reveals both notes to be in Petroff's handwriting.

58. Theodore C. Hinckley and Caryl Hinckley, "Ivan Petroff's Journal of a Trip to Alaska in 1878, Edited, with an Introduction and Annotations," *Journal of the West* 5, no. 1 (January 1966): 25–70.

59. Richard A. Pierce, "New Light on Ivan Petroff, Historian of Alaska," *Pacific Northwest Quarterly*, January 1968.

60. Ivan Petroff, *Population and Resources of Alaska*, House Executive Document 40, 4th Congress, 3d session, 1881.

61. Ivan Petroff, *Report on Population, Industries and Resources of Alaska* (Washington, DC: U.S. Government Printing Office, 1882).

62. Ernest Gruening, "Introduction," in Bancroft, *History of Alaska*, i–viii.

63. The map in the 1881 preliminary report is entitled "Sketch Map of Alaska Showing the Location of Settlements, Routes of Travel, Distances etc. to Accompany the Report of Ivan Petroff, Special Agent of the 10th Census." It includes most of the same names for the Yukon and Tanana tributaries as the Kandik Map. One of the maps published in the longer 1882 *Report*, and the same version republished in 1884, is entitled "Map of Alaska and Adjoining Regions compiled by Ivan Petroff, 1880." It shows the same tributaries with the same names, with some variations in spelling possibly arising from a misreading of Mercier's handwriting by later cartographers. It is available as part of the Online Collections, "Meeting of Frontiers" rare maps Web site of the Alaska & Polar Regions Collections, Rasmuson Library, University of Alaska Fairbanks. The map in the Bancroft *History of Alaska* is entitled "Map of Alaska and Adjoining Regions Compiled by Ivan Petroff, Special Agent Tenth Census." It includes the tributaries only on the east bank of the Yukon River and all without names.

64. It is possible that additional information on the sources and production of the map published in Bancroft's *History* might be located in the voluminous Bancroft Collection at the Bancroft Library.

65. George Davidson, "Explanation of an Indian Map of the Rivers, Lakes, Trails and Mountains from the Chilkaht to the Yukon Drawn by Chilkaht Chief Kohklux, in 1869," *Mazama* (April 1901). Article reprinted in *The Kohklux Map* (Whitehorse, YT: Yukon Historical and Museums Association, 1995).

66. John M. Campbell, *North Alaska Chronicle: Notes from the End of Time* (Santa Fe, NM: Museum of New Mexico Press, 1998).

67. Ogilvie, *Early Days*, 87. Numerous early visitors to the North carried published maps or sketches with them, including prospector Arthur Harper, as noted by Ogilvie and many others.

68. Mishler and Simeone, *Hän Hwëch'in*, 1–83.

69. Osgood, *The Han Indians*, 4.

70. Willie Juneby, "Place Names of the Eagle Region" (unpublished manuscript transcribed by John Ritter, August 1978, typescript at Yukon Native Language Centre, Whitehorse, YT), 76.

71. William Ogilvie, *Exploratory Survey of the Lewes, Tat-on-duc, Porcupine, Bell, Trout, Peel and Mackenzie Rivers 1887–88* (Ottawa, ON: Brown Chamberlain Printer to the Queen's Most Excellent Majesty, 1890), 55.

72. Leroy N. McQuesten, *Recollections of Life in the Yukon 1871–1885* (Dawson, YT: Yukon Order of Pioneers, 1952), 3.

73. Raboff, *Inuksuk Northern Koyukon*, 104.

74. Wickersham, *Old Yukon Tales*, 308.

75. Captain Charles Raymond, "Report of a Reconnaissance of the Yukon River, Alaska Territory. July to September, 1869," map (Washington, DC: U.S. Government Printing Office, 1871). It is available as part of the Online Collections, "Meeting of Frontiers" rare maps Web site of the Alaska & Polar Regions Collections, Rasmuson Library, University of Alaska Fairbanks.

76. Wickersham, *Old Yukon Tales*, 254.

77. Wright, *Prelude to a Bonanza*, 66, 110.

78. Mercier, *Recollections*, 66–9.

79. Dee Longenbaugh, "Alaska's Own Cartographers," *Terrae Incognitae* (1999): 61–9.

80. Ibid., 3.

81. Mishler and Simeone, *Hän Hwëch'in*, 94–106.

82. Marie-Francoise Guédon, *People of Tetlin, Why Are You Singing?* (Ottawa, ON: National Museums of Canada, 1974), xii, 8–16.

83. McQuesten, *Recollections of Life*, 6.

84. Allen, "Report of Lieut. Henry Allen," maps.

85. McQuesten, *Recollections of Life*, 8.

86. Schwatka, "Report of First Lieut. Frederick Schwatka," 340.

87. This was the spelling used in an interview given by Mercier years later, which mentioned "Katcé-Village" as one of three Indian communities in this area, named after the chief who died about 1880. Primum Picard, "Souvenirs de Voyages le Klondyke," *Le Monde Illustré* 14, no. 696 (September 4, 1897): 294, 297.

88. McQuesten, *Recollections of Life*, 4.

89. Allen, "Report of Lieut. Henry Allen," maps; Osgood, *The Han Indians*, 78.

90. Petroff, *Report on Population*, 69.

91. Ibid., 203.

92. Ferdinand Schmitter, *Upper Yukon Native Customs and Folk-lore* (Washington, DC: Smithsonian Institution, 1910), 16.

93. Jane Haigh, "... And His Native Wife" in *Preserving and Interpreting Cultural Change* (Anchorage: Alaska Historical Society Annual Meeting, 1996), 39–54.

94. Petroff, *Report on Population*, vi.

95. Dobrowolsky, *Hammerstones* 3–18.

96. Osgood, *The Han Indians*, 77–9.

97. Johnny Frank, *Johnny and Sarah Googwandak, Neerihiinjik We Traveled From Place to Place* (Fairbanks: Alaska Native Language Center, 1995).

98. Simon Paneak, *In a Hungry Country: Essays by Simon Paneak* (Fairbanks: University of Alaska Press, 2004), 59–68.

99. Raboff, *Inuksuk Northern Koyukon*, 117–51.

100. Ogilvie, *Exploratory Survey*, 55.

101. Julie Cruikshank, *Do Glaciers Listen? Local Knowledge, Colonial Encounters and Social Imagination* (Vancouver: University of British Columbia Press, 2005). Cruikshank's consideration of local knowledge and documentary sources contributed many ideas for the analysis of the Kandik Map in this chapter.

102. McQuesten, *Recollections of Life*, 8.

103. Petroff, *Population and Resources*, 62.

104. Ibid., 63.

105. Petroff, *Report on Population*, 257.

106. Ibid., 258.

107. Frederick Whymper, *Travel and Adventure in the Territory of Alaska* (London: John Murray, 1869), 226–9.

108. Alexander Hunter Murray, *Journal of the Yukon 1847–48* (Ottawa, ON: Government Printing Bureau, 1910), 34–6.

109. Robert McDonald, "Journals," in *Letters and Papers of the Church Missionary Society*, London (Typescript at Yukon Native Language Centre, Whitehorse, YT), September 10, 1865.

110. Reverend Vincent Sim, Journals, June 14, 1882; Annual Letter, January 19, 1883; Letter to Mr. Fenn, January 9, 1885. In *Letters and Papers of the Church Missionary Society*, London (Typescript at Yukon Native Language Centre, Whitehorse, YT).

111. Osgood, *The Han Indians*, 77–9.

112. Mercier, *Recollections*, 3.

113. Petroff, *Population and Resources*, 63.

114. McQuesten, *Recollections of Life*, 9.

115. Tappan Adney, "Moose Hunting with the Tro-chu-tin," *Harper's New Monthly Magazine* 100, no. 598 (March 1900).

116. Analysis provided by John Ritter, March 2007, Yukon Native Language Centre, Whitehorse, YT.

117. Juneby, "Place Names of the Eagle Region," 76.

118. Ibid., 76–8.

119. Ibid., 76–9.

120. Turck and Turck, "Trading Posts along the Yukon River," 51.

121. Jules Jette, *Koyukon Athabaskan Dictionary* (Fairbanks: Alaska Native Language Center, University of Alaska Fairbanks, 2000).

122. Haigh, ". . . And His Native Wife," 41–5.

123. Dr. Siri Tuttle, conversation with the author, Alaska Native Language Center, University of Alaska Fairbanks, August 2006.

124. John Ritter, conversation with the author, Yukon Native Language Centre, Whitehorse, YT, August 2006.

125. The "M" stands for Monsieur, not Moïse Mercier, who left the North in 1874.

126. Ivan Petroff to his wife, St. Michael, Alaska, July 8, 1880. Petroff Correspondence, The Bancroft Library, University of California, Berkeley.

127. McQuesten, *Recollections of Life*, 8.

128. Mercier, *Recollections*, 40.

129. Julie Cruikshank, *Reading Voices* (Vancouver, BC: Douglas & McIntyre, 1991), 1–21.

130. Murray, *Journal of the Yukon*, 82.

131. Turck and Turck, "Trading Posts along the Yukon River," 51–61.

132. Murray, *Journal of the Yukon*, 82.

133. Osgood, *The Han Indians*, 77.

134. Raymond, "Report of a Reconnaissance," 16–17.

135. William Ogilvie, *Information Respecting the Yukon District* (Ottawa, ON: Department of the Interior, 1898), 21.

136. William H. Dall, *Alaska and Its Resources* (New York: Arno/The New York Times, 1970), 105.

137. Ogilvie, *Information*, 27.

138. Whymper, *Travel*, 254. Whymper describes Red Leggings's colorful attire.

139. Mercier, *Recollections*, 54. Mercier wrote a whole section of his memoirs on Shahnyaati', spelled by him as Sénaté.

140. Linda Johnson, *An Index to the Journals of Reverend Robert McDonald* (Whitehorse, YT: Yukon Native Language Centre, 1984), B-1–B-312.

141. Ibid., B-103, B-163.

142. Jesuit Mission Records, MF no. 96, roll 21, Alaska & Polar Regions Collections, Rasmuson Library, University of Alaska Fairbanks.

143. Ogilvie, *Early Days*, 115.

144. Cruikshank, *Reading Voices*, 134.

145. Cruikshank and Robb, *Their Own Yukon*, 18, 57, 90.

146. Mishler and Simeone, *Hän Hwëch'in*, 256–70.

147. Arrowsmith, "British North America."

148. Dobrowolsky, *Hammerstones*, 3–16.

149. McClellan, *Part of the Land Part of the Water*, 206.

150. Johnson, *McDonald Index*, B-235.

151. Mishler and Simeone, *Hän Hwëch'in*, 256–70.

152. Ibid., 1–30.

153. Johnson, *McDonald Index*, B-73, B-177. Shahnyaati' was baptized as John Hardisty by Robert McDonald in 1866.

154. Dobrowolsky, *Hammerstones*, 77.

155. Ibid., 43–4.

156. Mishler and Simeone, *Hän Hwëch'in*, 12–13.

157. David Neufeld, "Gwitchin Visits to Fort Youcon 1847–1856" in *Notes from Your Historian* (Whitehorse, YT: Parks Canada, July 2004), 2–3.

158. Whymper, *Travel*, 257.

159. Robert McDonald, "Ven R. McDonald," manuscript life story transcribed by Rev. T. G. A. Wright, 1911, Robert McDonald Fonds, Yukon Archives, Whitehorse, YT.

160. McDonald, "Journals," June 1–4, 1866.

161. Ibid., November 18, 1866.

162. Ibid., March 12, 1867.

163. Whymper, *Travel*, 258.

164. Raymond, "Report of a Reconnaissance," 34.

165. Johnson, *McDonald Index*, B-15, B-177.

166. Neufeld, "Gwitchin Visits," 2–3.

167. Johnson, *McDonald Index*, B-18, B-104.

168. Ibid., A-1–A-12.

169. Molly Lee, "Context and Contact: The History and Activities of the Alaska Commercial Company, 1867–1900," in *Catalogue Raisonné of the Alaska Commercial Company Collection*, Nelson Graburn, Molly Lee, and Jen-Loup Rousselot, 19–20 (Berkeley: University of California Press, 1996). Lee notes that most of the A.C.C. records were destroyed in the fire that demolished the company offices after the San Francisco earthquake of 1906.

170. Mercier, *Recollections*, 54.

171. Ibid., 57.

172. Ibid., 43.

173. McDonald, "Journals," November 26, 1862; February 20, 1863; June 3, 1863.

174. Mercier, *Recollections*, 40.

175. Mercier, *Recollections*, 1.

176. McQuesten, *Recollections of Life*, 4.

177. Ibid., 7.

178. Ibid., 8.

179. Ibid., 5.

180. Unfortunately Harper was never able to find the promising prospect again on later trips.

181. Ibid.

182. Haigh, ". . . And His Native Wife," 41.

183. Ogilvie, *Early Days*, 101.

184. Ibid., 4–5.

185. Ibid., 5.

186. Ibid., 82.

187. Ivan Petroff to his wife, St. Michael's Redoubt, Alaska, July 8, 1880. Petroff Correspondence, The Bancroft Library, University of California, Berkeley.

188. Ibid.

189. Petroff, *Report on Population*, 161.

190. Ivan Petroff to his wife, Ikogmute, Yukon River, Alaska, August 2, 1880. Petroff Correspondence, The Bancroft Library, University of California, Berkeley.

191. Petroff, *Report on Population*, 60.

192. Schwatka, "Report of First Lieut. Frederick Schwatka," 293–301.

193. Frederick Schwatka, *Along Alaska's Great River* (Anchorage: Alaska Northwest Publishing Company, 1983), 30–1.

194. Schwatka, "Report of First Lieut. Frederick Schwatka," 340.

195. Schwatka's rendering of "Tthee t'äwdlenn", the Hän name for Eagle Bluff, sometimes written as "Fetolin" or "Tatotlin" by other early travelers.

196. Ibid.

197. Ibid., 343.

198. Ibid., 344.

199. Ibid., 345.

200. Allen, "Report of Lieut. Henry Allen," 437–42.

201. Ibid., 458–70.

202. McDonald, "Journals," July 1887.

203. Ogilvie, *Exploratory Survey*. The accompanying map contained details supplied by various Native people about the places along the route traveled by Ogilvie with them from the Yukon River to Fort McPherson.

204. W. F. Sparks, *A Friend to Man: The Story of Frank Sparks* (Victoria, BC: W. F. Sparks, 2006).

205. Green, *The Boundary Hunters*, 28–36.

206. McGrath Monthly reports, RG 32.

207. Willis V. Kingsbury, "Yukon River Diary," Vols. 1 and 2, 1889–1891, transcribed by Jim Paull, February 2008. Copy in personal collection of the author.

208. Photograph Albums of the Alaska Boundary Survey Party Presented to George Davidson by Dr. Kingsberry [*sic*], Davidson Collection 1946.006 Alb, The Bancroft Library, University of California, Berkeley. A second album contained photographs taken by the Turner U.S.C.&G.S. party on the Porcupine River.

209. Ibid., photograph no. 17, "A Band of David's Indians who camped about 3 1/2 miles above Camp Davidson during the winter of 1890–91."

210. Kingsbury, "Yukon River Diary," July 14–August 19, 1889.

211. Ibid., August 26–September 27, 1889.

212. Ibid., September 28–December 10, 1889.

213. Ibid., November 10–November 28, 1889.

214. Ibid., December 3, 1889–March 26, 1890.

215. Ibid., March 27–29, 1890.

216. Ibid., March 30, 1890.

217. Ibid.

218. Ibid., April 6–8, 1890.

219. Ibid., May 5–June 8, 1890.

220. Ibid., June 10–15, 1890.

221. Ibid., July 1–October 12, 1890.

222. Photograph Albums, Davidson Collection 1946.006 Alb, The Bancroft Library, photograph no. 18, "An Indian family and brush house."

223. Kingsbury, "Yukon River Diary," December 1, 1890–June 22, 1891.

224. Twelfth Census of the United States, 1900, MF roll 17 and 18, Alaska & Polar Regions Collections, Rasmuson Library, University of Alaska Fairbanks.

225. Canada. Statistics Canada. Census, Rampart House, 1891. MF68, reel no. 1–2. Yukon Archives, Whitehorse, YT.

226. Canada. Statistics Canada. Census, Yukon Territory, 1901. MF81, reel no. 1–3. Yukon Archives, Whitehorse, YT.

227. P. H. Ray and W. P. Richardson, "Report of Capt. P. H. Ray, Eighth Infantry, U.S.A. and Lieut. W. P. Richardson, Eighth Infantry, U.S.A.: 'Relief of the Destitute in the Yukon Region, 1898,'" in *Compilation of Narratives of Exploration in Alaska*, Senate Committee on Military Affairs, Senate Report 1023, 56th Congress, 1st session, 1900 (Washington, DC: U.S. Government Printing Office, 1900), 520.

228. Ibid., 534.

229. Ibid.

230. Linda Johnson, *Index to the Diaries of Archdeacon T. H. Canham* (Whitehorse, YT: Yukon Native Language Centre, 1988), 2.

231. Canada. Statistics Canada. Census, Yukon Territory, 1901, MF81, reel no. 1–3. Yukon Archives, Whitehorse, YT.

232. William Schneider, "Chief Sesui and Lieutenant Herron: A Story of Who Controls the Bacon," in *An Alaska Anthology: Interpreting the Past*, ed. Stephen W. Haycox and Mary Childers Mangusso (Seattle: University of Washington Press, 1996), 176–90.

233. Mercier, *Recollections*.

234. Québec Bibliothèque et Archives nationales, Revues d'un siècle, http://bibnum2.banq.qc.ca/bna/illustrations/accuiel.

235. Jesuit Mission Records, MF no. 96, roll 21, Frame 58, Nulato, 1875. Alaska & Polar Regions Collections, Rasmuson Library, University of Alaska Fairbanks.

236. Johnson, *McDonald Index*, c–31.

237. Picard, "Souvenirs de Voyages," 294.

238. Élisée Reclus, *Nouvelle Géographie Universelle La Terre et Les Hommes xv Amérique Boréale* (Paris: Librarie Hachette et Cie., 1890), 208.

239. "Frs. Mercier, Célèbre Voyageur Canadien," *L'opinion publique* 2, no. 43 (October 26, 1871): 517. Québec Bibliothèque et Archives nationales, Revues d'un siecle, http://bibnum2.banq.qc.ca/bna/illustrations/accuiel.

240. "Scènes du Klondyke: Station Mercier (Station Tanana)," *Le Monde Illustré* 14, no. 696 (September 4, 1897): 297. Québec Bibliothèque et Archives nationales, Revues d'un siecle, http://bibnum2.banq.qc.ca/bna/illustrations/accuiel.

241. Wickersham, *Old Yukon Tales*, 159.

242. Jesuit Mission Records, MF no. 96, roll 21, Frame 58, Nulato, 1875. Alaska & Polar Regions Collections, Rasmuson Library, University of Alaska Fairbanks.

243. Melody Webb, *The Last Frontier: A History of the Yukon Basin of Canada and Alaska* (Albuquerque: University of New Mexico Press, 1985), 56–75.

244. In his *Géographie*, Reclus places Mercier's name along the mid Yukon, Tanana, and Kuskokwim on a map labeled "Principaux Itinéraires des voyageurs dans l'Alaska," 195. Neither Mercier nor any of his colleagues such as McQuesten mention him traveling on the Tanana and Kuskokwim.

245. Lee, "Context and Contact," 19–20.

246. Mishler and Simeone, *Hän Hwëch'in*, 256–69.

247. Elva R. Scott, *Jewel on the Yukon: Eagle City Collection of Essays on Historic Eagle and Its People* (Eagle City, AK: Eagle Historical Society and Museums, 1997), 117. This history does not clarify whether the church was named in honor of François Xavier Mercier's namesake saint.

248. Leon Trepanier, "L'Aventureuse Carrière de Moïse Mercier, de St-Paul L'Ermite, le Premier de Race Blanche à Explorer les Rivières de l'Alaska," *La Patrie*, October 1, 1950.

249. André Mercier, conversation with the author, Gatineau, Québec, June 2006.

250. Ibid.

251. "Elections municipales, Village propre, M. Moise Mercier, Maire de St. Véronique de Turgeon," unidentified newspaper clipping, circa 1907.

252. Yann Herry, *La Francophonie Une richesse nordique Northern Portraits* (Whitehorse, YT: L'Association franco-yukonnaise, 2004), 26–7.

253. O. L. David, "Galerie Nationale François Mercier," *L'opinion publique* 2, no. 43 (November 2, 1871): 517, 525.

254. Ibid.

255. Ibid.

256. Ibid. The article included a misprint, citing 1869 as the year of Mercier's first trip to the North.

257. Ibid.

258. Ibid.

259. Ibid.

260. Hector Chevigny, *The Great Alaskan Venture 1741–1867* (Portland, OR: Binford & Mort, 1965).

261. Dee Longenbaugh, "Alaska's Own Cartographers," *Terrae Incognitae* 31 (1999): 61–69.

262. Jennifer S. H. Brown, *Strangers in Blood: Fur Trade Company Families in Indian Country* (Vancouver: University of British Columbia Press, 1980).

263. Wright, *Prelude to a Bonanza,* 1–117.

264. Raboff, *Inuksuk Northern Koyukon,* 115–63.

265. Murray, *Journal of the Yukon,* 28.

266. Ibid., 74.

267. Wright, *Prelude to a Bonanza,* 70–3.

268. McDonald, "Journals," 1862–63.

269. Patrick Moore, "Archdeacon Robert McDonald and Tukudh (Gwich'in) Literacy," *Anthropological Linguistics* (March 2008). Copy in personal collection of Linda Johnson.

270. McDonald, "Journals," June 5, 1866.

271. Dall, *Alaska and Its Resources,* 53.

272. Ibid., 110.

273. Ibid., 58.

274. Ibid., 102.

275. Ibid.

276. Ibid.

277. McDonald, "Annual Letter to the C.M.S. Secretaries," October 29, 1867.

278. McDonald, "Journals," August 1870.

279. Mercier, *Recollections,* 67.

280. Ibid., 66–73.

281. McDonald, "Journals," December 31, 1869.

282. Ibid., June 7, 1870.

283. Ibid., July 6–21, 1870.

284. Ibid., August 4, 1870.

285. Ibid.

286. Ibid., July 17, 1870.

287. Ibid., August 17, 1870.

288. Ibid., September 12, 1870.

289. McDonald, "Annual Letter," June 30, 1871.

290. McDonald, "Journals," July 4–31, 1871.

291. David, "François Mercier," 525.

292. McDonald, July 24, 1871.

293. Ibid.

294. Johnson, *McDonald Index,* A-6–A-7.

295. Kenneth McDonald, "Journals," February 10, 1875.

296. Mercier, *Recollections,* 12.

297. Ibid., 15.

298. Ibid., 12.

299. Ibid., 38.

300. Ibid., 40.

301. Robert McDonald, "Annual Letter," 1874.

302. Ed Jones and Star Jones, *All that Glitters: The Life and Times of Joe Ladue, Founder of Dawson City* (Whitehorse, YT: Wolf Creek Books, 2005), 56–7.

303. Mercier, *Recollections*, 26.

304. McQuesten, *Recollections of Life*, 2.

305. Michael Gates, *Gold at Fortymile Creek: Early Days in the Yukon* (Vancouver: University of British Columbia Press, 1994), 5–8.

306. McQuesten, *Recollections of Life*, 1.

307. Ibid., 3.

308. Ibid.

309. Ibid., 4.

310. Mercier, *Recollections*, 27–8.

311. Ibid., 28.

312. Ibid., 3.

313. Ibid.

314. Chief Catsah was likely the same man McDonald called Katza, and Mercier called Katcé.

315. McQuesten, *Recollections of Life*, 4.

316. Mercier, *Recollections*, 28.

317. McQuesten, *Recollections of Life*, 5.

318. Mercier, *Recollections*, 67.

319. McQuesten, *Recollections of Life*, 5.

320. Mercier apparently traveled on the lower part of that river at some point during his career, according to information he provided to Reclus, who published his *Géographie* in 1890 with manuscript notes supplied by Mercier.

321. Ogilvie, *Information*, 70.

322. Jesuit Mission Records, MF no. 96, roll 21, Frame 58, Nulato, 1875. Alaska & Polar Regions Collections, Rasmuson Library, University of Alaska Fairbanks.

323. Mercier, *Recollections*, 5.

324. McQuesten, *Recollections of Life*, 7.

325. Mercier, *Recollections*, 22.

326. McQuesten, *Recollections of Life*, 6.

327. Mercier, *Recollections*, 21. Mercier said that Harper made the trip with "three of our Indian servants," presumably meaning people he employed at his post, but no other details were given by either him or McQuesten, who also wrote that Harper retrieved Mrs. Bean's body.

328. McQuesten, *Recollections of Life*, 7.

329. Mercier gave its name in French as the *St. Michel* in his *Recollections*.

330. Ibid., 8.

331. Reclus, *Nouvelle Géographie*, 195.

332. McQuesten, *Recollections of Life*, 7.

333. Mercier, *Recollections*, 59. Like the rest of the traders Mercier may have been collecting specimens for scientists at the Smithsonian or for other agencies.

334. Theodore C. Hinckley and Caryl Hinckley, "Ivan Petroff's Journal of a Trip to Alaska in 1878, Edited, with an Introduction and Annotations," *Journal of the West* 5, no. 1 (January 1966): 49–51.

335. Mercier, *Recollections*, 61.

336. Petroff, *Population and Resources*, 86.

337. Petroff, *Report on Population*, 111.

338. Ibid.

339. Ibid.

340. Ibid., 160.

341. Ivan Petroff to his wife, St. Michael's Redoubt, Alaska, July 8, 1880, and August 2, 1880, Ikogomute, Alaska, Petroff Correspondence, The Bancroft Library, University of California, Berkeley.

342. Picard, "Souvenirs de Voyages," 294.

343. Mercier, *Recollections*, 3.

344. McQuesten, *Recollections of Life*, 9.

345. Mercier, *Recollections*, 61.

346. McQuesten, *Recollections of Life*, 10.

347. Mercier, *Recollections*, 3.

348. McQuesten, *Recollections of Life*, 10.

349. Ibid., 11.

350. Mercier, *Recollections*, 32.

351. McQuesten, *Recollections of Life*, 12.

352. Mercier, *Recollections*, 65.

353. "Scènes du Klondike: Station Mercier (Station Tanana)," *Le Monde Illustré* 14, no. 696 (September 4, 1897): 297. Québec Bibliothèque et Archives nationales, Revues d'un siècle, http://bibnum2.banq.qc.ca/bna/illustrations/accuiel.

354. Mercier, *Recollections*, 5.

355. McQuesten, *Recollections of Life*, 14.

356. Mercier, *Recollections*, 62.

357. Ibid., 70.

358. Lee, "Context and Contact," 24.

359. That loyalty continued for many years, as Ogilvie noted in his 1888 report, *Information,* 46–47: "David's and Charley's bands manifested to me a much stronger sympathy for Canada than for the United States . . . [perhaps] due to policy . . . but . . . most of their dealings and all their education have been Canadian."

360. "Scènes du Klondike," 294. The newspaper article identified the people in the photo: at bottom left, seated between two Indians, the Chief Starke Souhage; on the porch standing at the left, Androuska; Russian Métis; his wife Madronna; François Mercier, seated with his dog Jack beside him; standing beside the door, Sport, Chief of the Nowikaket Indians; J. Beaudoin, between the door and window; and Siroska, between the two windows.

361. Wickersham, *Old Yukon Tales*, 159. Wickersham identified the photo as taken in 1883, while Yarborough references it as 1885. The later date would appear to be correct,

as Willis Everette and Mercier left together that year. Charles Farciot was engineer on the New Racket, a small steamer brought up the Yukon by the Schieffelin prospecting group in 1883.

362. Ogilvie felt the Merciers had not been able to adjust to the trading preferences of the Yukon Indians, established during the H.B.C. days. Ogilvie, *Early Days*, 64.

363. Robert S. Coutts, *Yukon Places and Names* (Sidney, BC: Gray's Publishing Limited, 1980), 121, 174, 180–1.

364. Wickersham, *Old Yukon Tales*, 150.

365. Jones and Jones, *All that Glitters*, 107–227.

366. Reclus, *Nouvelle Géographie*, 210. Reclus wrote for example that Mercier had ascended the Tanana, "the first among the whites" in 1848, probably a misprint of 1878 or 1884, but still not the first trip recorded up the Tanana by that time. On page 256 he stated that Mercier had established Fort Reliance for the H.B.C.

367. Mercier, *Recollections*, xii.

368. *Unveiling the Arctic*, ed. Louis Rey (Calgary, AB: The Arctic Institute of North America, 1984).

369. Wright, *Prelude to a Bonanza*, 1–117.

370. Longenbaugh, "Alaska's Own Cartographers," 61–9.

371. Marvin Falk, "European Images of Pre-discovery Alaska," in *Unveiling the Arctic*, 562–73.

372. Wright, *Prelude to a Bonanza*, 9.

373. The map was entitled "A New and Accurate Map of North America Including Nootka Sound; with the New Discovered Islands on the Northeast Coast of Asia." Rare Map G3300 [1780] C66, Alaska & Polar Regions Collections, Rasmuson Library, University of Alaska Fairbanks. It is available as part of the Online Collections, "Meeting of Frontiers" rare maps Web site of the Alaska & Polar Regions Collections, Rasmuson Library, University of Alaska Fairbanks.

374. M. D. Teben'kov, *Atlas of the Northwest Coast of America*, trans. and ed. R. A. Pierce (Kingston, ON: The Limestone Press, 1981), 4–6.

375. Various spellings of this name appear in different early sources. See Donald J. Orth, *Dictionary of Alaska Place Names* (Washington, DC: U.S. Government Printing Office, 1971), 478.

376. Henry N. Michael, *Lieutenant Zagoskin's Travels in Russian America, 1842–1844* (Toronto, ON: University of Toronto Press, 1967), 84.

377. The map was entitled *Karta Rossiiskikh Baabhih*. Rare Map G4370 1861 K37, Alaska & Polar Regions Collections, Rasmuson Library, University of Alaska Fairbanks.

378. Longenbaugh, "Alaska's Own Cartographers," 61–9.

379. U.S. Coast and Geodetic Survey, *United States Coast Pilot 8: Pacific Coast. Alaska. Dixon Entrance to Cape Spencer* (Washington, DC: U.S. Government Printing Office, 1869).

380. Wright, *Prelude to a Bonanza*, 16.

381. Ibid., 49.

382. Ibid., 70–7.

383. Barbara Belyea, "Inland Journeys, Native Maps," *Cartographica* 33, no. 2 (Summer 1996).

384. George Dawson, *Report of an Exploration in the Yukon District, N.W.T. and Adjacent Northern Portion of British Columbia 1887* (Whitehorse, YT: Yukon Historical and Museums Association, 1987), 138.

385. Arrowsmith, "British North America."

386. Murray, *Journal of the Yukon*, 74. See also Richardson's map included in this report, facing page 74.

387. Ibid., 73.

388. Ibid., 74.

389. Ibid., 78.

390. Hardisty Map, 1853, in Raboff, *Inuksuk Northern Koyukon*, 97.

391. James McDougall, "Sketch Map of Yukon River," C38/25 fo. 54 (N11325), Hudson's Bay Company Archives, Manitoba Archives, Winnipeg, MB.

392. Raymond, "Report of a Reconnaissance," 8.

393. Alaska and Yukon Territory, 1869, Rare Map G4370 1869 U55, Alaska & Polar Regions Collections, Rasmuson Library, University of Alaska Fairbanks.

394. Charles Raymond, "The Yukon River, Alaska from Fort Yukon to the Sea." Rare Map digital ID f87101 http://hdl.loc.gov/loc.ndlpcoop/mtfxmp.f87101, Alaska & Polar Regions Collections, Rasmuson Library, University of Alaska Fairbanks. It is available as part of the Online Collections, "Meeting of Frontiers" rare maps Web site of the Alaska & Polar Regions Collections, Rasmuson Library, University of Alaska Fairbanks.

395. Oscar Lewis, *George Davidson: Pioneer West Coast Scientist* (Berkeley: University of California Press, 1954), 40–52.

396. "The Kohklux Map," George Davidson Collection, Map G4370 1852 K6, The Bancroft Library, University of California, Berkeley.

397. Robert Campbell, *Journal of Occurrences at the Forks of the Lewes and Pelly Rivers, May 1848 to September 1852,* transcribed and edited with notes by Llewellyn Johnson (Whitehorse, YT: Yukon Government Heritage Branch, 1995).

398. Davidson, "Explanation of an Indian Map," 16.

399. A note on the Kohklux map reads "Loaned by Prof Davidson To Be Returned."

400. Davidson, "Explanation of an Indian Map," 14–24.

401. Ibid., 16.

402. Raymond, "Report of a Reconnaissance," 35.

403. Ibid., 24.

404. Ibid., 8.

405. Dall used an English rendition, LeBarge. The official toponym is spelled Laberge, matching the francophone origins of Michel.

406. Coutts, *Yukon Places and Names*, 151.

407. Ruth Gotthardt, *Ta'an Kwach'an People of the Lake* (Whitehorse, YT: Government of the Yukon, 2000), 2.

408. Whymper, *Travel*, map published with book.

409. McDonald, "Journals," September 15, 1870.

410. Wright, *Prelude to a Bonanza*, 86–91.

411. William Carpenter Bompas, *Diocese of MacKenzie River* (London: Society for Promoting Christian Knowledge, New York: E. & J.B. Young, 1888), 39. Yukon Archives Pamphlet Collection, PAM 1888-0009. Bishop Bompas reported that "The Tukudh tribes have a national tradition of having reached their present country by crossing an icy strait of the sea, which was probably Behring's Strait."

412. Ogilvie, *Early Days*, 88.

413. Ibid.

414. Ibid., 100.

415. Ibid.

416. McQuesten, *Recollections of Life*, 5.

417. Ogilvie, *Early Days*, 101.

418. Ibid.

419. Ibid.,165.

420. Petroff, *Report on Population*, vi.

421. Ibid., 81.

422. Aurel Krause and Arthur Krause, *Journey to the Tlingits 1881/82*, translated by Margot Krause McCaffrey (Haines, AK: Haines Centennial Commission, 1981), 27.

423. Petroff, *Report on Population*, 83.

424. Ibid., 82.

425. Petroff Correspondence, Courtesy of The Bancroft Library, University of California, Berkeley.

426. Ibid.

427. Belle Herbert, *Shandaa, In My Lifetime* (Fairbanks: Alaska Native Language Center, 1982), 4.

428. Osgood, *The Han Indians*, 14.

429. Donald W. Clark, *Fort Reliance, Yukon: An Archaeological Assessment* (Ottawa, ON: Canadian Museum of Civilization, 1995), 35.

430. Johnson, Manuscript Map H2/600/1888 NMC023871, Map Collection, Library & Archives Canada, Ottawa, ON.

431. British War Office, Map of Alaska and Adjoining Regions, Map H3/1209/Alaska/1886 NMC 0086698, Map Collection, Library & Archives Canada, Ottawa, ON.

432. Green, *The Boundary Hunters*, 19–45.

433. Tim Ingold, *The Perception of the Environment* (London: Routledge, 2000), 219.

434. Ibid.

435. Ibid., 220.

436. Ibid.

437. Ibid., 223.

438. Ibid., 225.

439. Ibid., 226.

440. Ibid., 227.

441. Ibid., 228.

442. Ibid., 231.

443. Keith Basso, "'Stalking with Stories': Names, Places, and Moral Narratives among the Western Apache," in *Text, Play, and Story: The Construction and Reconstruction of Self and Society*, ed. Edward M. Bruner (Washington, DC: American Ethnological Society, 1984), 26.

444. Ibid.

445. Ibid., 25.

446. Ibid., 27

447. Ibid.

448. Ibid., 45–6. Also cited in Robert Rundstrom, "The Role of Ethics, Mapping and the Meaning of Place in Relations Between Indians and Whites in the United States."

449. Ibid., 50.

450. Ryden, *Mapping the Invisible Landscape*, 1.

451. Ibid., 114.

452. Ibid., 98.

453. Ibid., 114.

454. There are two Kohklux Maps. Chief Kohklux drew the first map for Davidson on his own while he was in jail at Sitka. This map is smaller than the second map drawn at Klukwan. It was probably not influenced as much by western cartography, nor did it include the knowledge of his wives. It is less detailed, and presents the route to Fort Selkirk as a circular pattern of travel. Davidson annotated the map later: "Kohklux started from his place at Klukwan and drew all around the paper for want of room." Subsequently Davidson drew his own version of the Tlingit routes in 1893 that incorporated details from the second, larger Kohklux Map and information from explorers such as Schwatka, Ogilvie, Dalton, and Glave, who traveled in the region after Kohklux. This map was published with his *Mazama* article in 1901.

455. Davidson, "Explanation," in *The Kohklux Map*, 16.

456. Lewis, *George Davidson*, 90.

457. Campbell, 26.

458. Ibid., 101–2.

459. Katherine Peter, *Neets'aii Gwindaii. Living in the Chandalar Country*, retranslated by Adeline Raboff (Fairbanks: Alaska Native Language Center, 1992), 101–3.

460. Ibid., viii–x.

461. Ibid., 100.

462. Belyea, "Inland Journeys."

463. Nicolas Peterson and Marcia Langdon, *Aborigines, Land and Land Rights* (Canberra: Australian Institute of Aboriginal Studies, 1983), 118–29.

464. Linda Finn Yarborough, conversation with the author, Lake Laberge, circa July 1984. The Kandik Map is reprinted on the inside of the back cover of Mercier, *Recollections*.

465. Percy Henry, conversation with the author, Yukon Native Language Centre, Whitehorse, YT, April 1, 2004.

466. Gerald Isaac, "The Han Huch'inn Early Warning System," 1.

467. Gerald Isaac, "1880 Kandik Map," presentation at Upper Yukon River Heritage Symposium, Dawson City, YT, March 8–10, 2001.

468. Georgette McLeod, Tr'ondëk Hwëch'in Heritage Department, telephone conversation with the author, Dawson City and Whitehorse, YT, December 2007.

469. Effie Kokrine, conversation with the author, Fairbanks, October 2001.

470. Phyllis A. Fast, *Weavers of Two Worlds: Athabascan Women of the Gold Rush Era* (Fairbanks: University of Alaska Museum, n.d.).

471. Phyllis A. Fast, conversation with the author, Fairbanks, November 2001.

472. André Mercier, conversation with the author, Gatineau, Québec, June 2006.

473. Bessie Johns, "Mrs. Bessie Johns' Account of the Border Survey," 1989 Y.H.M.A. Borderlands Conference in Easton, *An Ethnohistory of the Chisana River Basin* (Whitehorse, YT: Northern Research Institute, Yukon College, 2006), 154–7.

474. Lulla Sierra John, conversation with the author, Vancouver, November 2006.

475. Gerald Isaac, conversation with the author, Whitehorse, August 2006.

476. McClellan, *Part of the Land Part of the Water.*

477. Isaac Juneby, conversation with the author, Eagle Village, August 2006.

478. Andy Bassich, conversation with the author, on board the *Yukon Queen II*, Yukon River, September 2006.

479. Discovering Northern Gold Symposium, Yukon Historical and Museums Association, Whitehorse, YT, October 2006. Program at http://www.yukonalaska.com/yhma/ngold.pdf.

480. Hän Literacy Session, Yukon Native Language Centre, Whitehorse, YT, November 2006.

481. Yann Herry, conversation with the author, Whitehorse, November 2006.

482. David Neufeld, conversation with the author, Whitehorse, November 2006.

483. Adeline Raboff, conversation with the author, Fairbanks, November 2006.

Selected Bibliography

Adney, Tappan. "Moose Hunting with the Tro-chu-tin." *Harper's New Monthly Magazine* 100, no. 598 (March 1900): 494–507.

Anonymous. "Elections municipales, Village propre, M. Moise Mercier, Maire de St. Véronique de Turgeon." Unidentified newspaper clipping, circa 1907. Copy in personal collection of Linda Johnson.

Arrowsmith, John. *British North America.* London: J. Arrowsmith, 1854. Reproduction of original map in the Map Library, University of Western Ontario. Ottawa and Whitehorse: Association of Canadian Map Libraries and Yukon Historical and Museums Association, 1982.

Bancroft, Hubert H. *History of Alaska 1730–1885.* New York: Antiquarian Press Ltd., 1959.

Basso, Keith. "Stalking with Stories: Names, Places, and Moral Narratives among the Western Apache." In *Text, Play, and Story: the Construction and Reconstruction of Self and Society*, edited by Edward M. Bruner. Washington, DC: American Ethnological Society, c. 1984.

Berger, Thomas R. *Northern Frontier Northern Homeland: The Report of the Mackenzie Valley Pipeline Inquiry.* Vol. 1. Ottawa, ON: Department of Supply and Services, 1977.

———. *Village Journey: The Report of the Alaska Native Review Commission.* New York: Hill and Wang, 1985.

Berton, Pierre. *Klondike, The Last Great Gold Rush.* Toronto, ON: McClelland & Stewart Limited, 1972.

Bompas, William Carpenter. *Diocese of MacKenzie River.* London: Society for Promoting Christian Knowledge, New York: E. & J. B. Young, 1888. Yukon Archives Pamphlet Collection, PAM 1888-0009.

Brooks, Alfred Hulse. *Blazing Alaska's Trails.* Fairbanks and Washington, DC: University of Alaska and Arctic Institute of North America, 1953.

Brown, Jennifer S. H. *Strangers in Blood: Fur Trade Company Families in Indian Country.* Vancouver: University of British Columbia Press, 1980.

Campbell, John M. *North Alaska Chronicle: Notes from the End of Time*. Santa Fe: Museum of New Mexico Press, 1998.

Campbell, Robert. *Journal of Occurrences at the Forks of the Lewes and Pelly Rivers*. Edited by Llewellyn R. Johnson. Whitehorse, YT: Government of Yukon, 1995.

Chevigny, Hector. *The Great Alaskan Venture 1741–1867*. Portland, OR: Binford & Mort, 1965.

Clark, Donald W. *Fort Reliance, Yukon: An Archaeological Assessment*. Ottawa, ON: Canadian Museum of Civilization, 1995.

Coates, Ken S., and William R. Morrison. *Land of the Midnight Sun: A History of the Yukon*. Edmonton, AB: Hurtig Publishers, 1988.

Cody, H. A. *An Apostle of the North: Memoirs of the Right Reverend William Carpenter Bompas, D. D.* Toronto, ON: The Musson Book Co. Limited, 1908.

Cooke, C. "A New and Accurate Map of North America Including Nootka Sound; with the New Discovered Islands on the Northeast Coast of Asia." Fairbanks, AK: Rasmuson Library Rare Map (G3300 [1780] C66).

Coutts, Robert S. *Yukon Places and Names*. Sidney, BC: Gray's Publishing Limited, 1980.

Cruikshank, Julie. *Life Lived Like a Story*. Vancouver: University of British Columbia Press, 1990.

_______. *Reading Voices. Dan Dha Ts'edenintth'e. Oral and Written Interpretations of the Yukon's Past*. Vancouver, BC: Douglas & McIntyre, 1991.

_______. *The Social Life of Stories: Narrative and Knowledge in the Yukon Territory*. Vancouver: University of British Columbia Press, 1998.

_______. *Do Glaciers Listen? Local Knowledge, Colonial Encounters and Social Imagination*. Vancouver: University of British Columbia Press, 2005.

Cruikshank, Julie, and Jim Robb. *Their Own Yukon: A Photographic History by Yukon Indian People*. Whitehorse, YT: Yukon Indian Cultural Education Society and Yukon Native Brotherhood, 1975.

Dall, William H. *Alaska and Its Resources*. New York: Arno/The New York Times, 1970.

David, O. L. "Galerie Nationale François Mercier." *L'opinion publique* 2, no. 43 (November 2, 1871): 517.

Davidson, George. "Explanation of an Indian Map of the Rivers, Lakes, Trails and Mountains from the Chilkaht to the Yukon Drawn by Chilkaht Chief Kohklux, in 1869." *Mazama*, April 1901.

Dawson, George. *Report of an Exploration in the Yukon District, N.W.T. and Adjacent Northern Portion of British Columbia 1887*. Montreal, QC: William Foster Brown & Co., 1889. Reprint, Whitehorse, YT: Yukon Historical and Museums Association, 1987.

Dobrowolsky, Helen. *Hammerstones: A History of the Tr'ondëk Hwëch'in*. Dawson City, YT: Tr'ondëk Hwëch'in, 2003.

Easton, N. A. *An Ethnohistory of the Chisana River Basin*. Whitehorse, YT: Northern Research Institute, Yukon College, 2006.

Falk, Marvin. "European Images of Pre-discovery Alaska." In *Unveiling the Arctic*, edited by Louis Rey. Calgary, AB: The Arctic Institute of North America, 1984.

———. *Alaska World Bibliographic Series*. Vol. 183. Oxford: Clio Press, 1995.

Fast, Phyllis A. *Weavers of Two Worlds: Athabascan Women of the Gold Rush Era*. Fairbanks: University of Alaska Museum, n.d.

Ferguson, Judy. *Parallel Destinies: An Alaskan Odyssey*. Big Delta, AK: Glas Publishing Company, 2002.

Gates, Michael. *Gold at Fortymile Creek: Early Days in the Yukon*. Vancouver: University of British Columbia Press, 1994.

Geddes, Carol. *Picturing a People: George Johnston, Tlingit Photographer*, videocassette. Montreal, QC: National Film Board of Canada in co-production with Nutaaq Média Inc. and Fox Point Productions, 1997.

Girard, Cécile, and Renée Laroche. *Un Jardin sur le Toit: La Petit Histoire des Francophones du Yukon*. Whitehorse, YT: L'Association franco-yukonnaise, 1991.

Gotthardt, Ruth. *Ta'an Kwach'an: People of the Lake*. Whitehorse, YT: Government of the Yukon, 2000.

Graburn, Nelson, Molly Lee, and Jen-Loup Rousselot, *Catalogue Raisonné of the Alaska Commercial Company Collection*. Berkeley: University of California Press, 1996.

Green, Lewis. *The Boundary Hunters: Surveying the 141st Meridian and the Alaska Panhandle*. Vancouver: University of British Columbia Press, 1982.

Guédon, Marie-Françoise. *People of Tetlin, Why Are You Singing?* Ottawa, ON: National Museums of Canada, 1974.

Haigh, Jane. ". . . And His Native Wife." In *Preserving and Interpreting Cultural Change*. Anchorage: Alaska Historical Society Annual Meeting, 1996.

Herbert, Belle. *Shandaa, In My Lifetime*. Fairbanks: Alaska Native Language Center, 1982.

Herry, Yann. *La Francophonie Une Richesse Nordique Northern Portraits*. Whitehorse, YT: L'Association franco-yukonnaise, 2004.

Hinckley, Theodore C., and Caryl Hinckley. "Ivan Petroff's Journal of a Trip to Alaska in 1878, Edited, with an Introduction and Annotations." *Journal of the West*, 5, no. 1 (January 1966): 25–70.

Ingold, Tim. *The Perception of the Environment*. London: Routledge, 2000.

Isaac, Gerald. "1880 Kandik Map." Paper presented at the Upper Yukon River Heritage Symposium, Dawson City, Yukon, March 8–10, 2001.

Jette, Jules. *Koyukon Athabaskan Dictionary*. Fairbanks: Alaska Native Language Center, 2000.

John, Katie. *Tatl'ahwt'aenn Nenn': The Headwaters People's Country: Narratives of the Upper Ahtna Athabaskans*. Transcribed and edited by James Kari. Fairbanks: Alaska Native Language Center, 1986.

Johnson, Linda. "The Day the Sun Was Sick." *Yukon Indian News*, Summer 1984. Whitehorse, YT: Ye-Sa-To Communications Society.

———. *An Index to the Journals of Reverend Robert McDonald 1862–1913*. Whitehorse, YT: Yukon Native Language Centre, 1985.

———. *An Index to the Diaries of Archdeacon T. H. Canham 1899–1915.* Whitehorse, YT: Yukon Native Language Centre, 1988.
Jones, Ed, and Star Jones. *All that Glitters: The Life and Times of Joseph Ladue, Founder of Dawson City.* Whitehorse, YT: Wolf Creek Books, 2005.
Krause, Aurel, and Arthur Krause. *Journey to the Tlingits 1881/82.* Translated by Margot Krause McCaffrey. Haines, AK: Haines Centennial Commission, 1981.
Lewis, Oscar. *George Davidson: Pioneer West Coast Scientist.* Berkeley: University of California Press, 1954.
Longenbaugh, Dee. "Alaska's Own Cartographers." *Terrae Incognitae,* 31 (1999): 61–9.
MacDonald, Linda E. T., and Lynette R. Bleiler. *Gold & Galena: A History of the Mayo District.* Mayo, YT: Mayo Historical Society, 1990.
McClellan, Catharine. "Culture Change and Native Trade in Southern Yukon Territory." Ph.D. diss., University of California, Berkeley, 1950.
———. "Wealth Woman and Frogs among the Tagish Indians." *Anthropos* 58 (1963): 121–8.
———. "Culture Contacts in the Early Historic Period in Northwestern North America." *Arctic Anthropology* 2 (1964): 3–15.
———. *Part of the Land Part of the Water: A History of the Yukon Indians.* Vancouver, BC: Douglas & McIntyre, 1987.
———. "Before Boundaries, People of Yukon/Alaska." Paper presented at the Yukon Historical and Museums Association Conference, Whitehorse, June 2–4, 1989. http://www.yukonalaska.com/yhma/public/mcclell.htm.
McQuesten, Leroy N. *Recollections of Life in the Yukon 1871–1885.* Dawson, YT: Yukon Order of Pioneers, 1952.
Mercier, François Xavier. *Recollections of the Youkon: Memoires from the Years 1868–1885.* Edited by Linda Finn Yarborough. Anchorage: The Alaska Historical Society, 1986.
Michael, Henry N. *Lieutenant Zagoskin's Travels in Russian America, 1842–1844.* Toronto, ON: University of Toronto Press, 1967.
Mishler, Craig, ed. *Johnny Sarah Hàa Googwandak. Neerihiinjìk. We Traveled from Place to Place. The Gwich'in Stories of Johnny and Sarah Frank.* Fairbanks: Alaska Native Language Center, 1995.
Mishler, Craig, and William E. Simeone. *Hän Hwëch'in: People of the River.* Fairbanks: University of Alaska Press, 2004.
Murray, Alexander Hunter. *Journal of the Yukon 1847–48.* Ottawa, ON: Government Printing Bureau, 1910.
Neufeld, David. "Parks Canada and the Commemoration of the North: History and Heritage." In *Northern Visions: New Perspectives on the North in Canadian History,* edited by Kerry Abel and Ken S. Coates, 45–75. Peterborough, ON: Broadview Press, c. 2001.
———. "The Commemoration of Northern Aboriginal Peoples by the Canadian Government." *The George Wright Forum* 19, no. 3 (2002): 22–33.
———. "Gwitchin Visits to Fort Youcon 1847–1856." In *Notes from Your Historian.* Whitehorse, YT: Parks Canada, July 2004.

Ogilvie, William. *Exploratory Survey of the Lewes, Tat-on-duc, Porcupine, Bell, Trout, Peel and Mackenzie Rivers 1887–88*. Ottawa, ON: Brown Chamberlain Printer to the Queen's Most Excellent Majesty 1890.

———. *Information Respecting the Yukon District*. Ottawa, ON: Department of the Interior, 1898.

———. *Early Days on the Yukon*. Ottawa, ON: Thornburn & Abbott, 1913.

Orth, Donald, J. *Dictionary of Alaska Place Names Geological Survey Professional Paper 567*. Washington, DC: U.S. Government Printing Office, 1971.

Osgood, Cornelius. *Contributions to the Ethnography of the Kutchin*. New Haven, CT: Yale University Press, 1970.

———. *The Han Indians: A Compilation of Ethnographic and Historical Data on the Alaska-Yukon Boundary Area*. New Haven, CT: Yale University Press, 1971.

Paneak, Simon. *In a Hungry Country: Essays by Simon Paneak*. Fairbanks: University of Alaska Press, 2004.

Peter, Katherine. *Neets'aii Gwindaii. Living in the Chandalar Country*. Retranslated by Adeline Raboff. Fairbanks: Alaska Native Language Center, 1992.

Peterson, Nicolas, and Marcia Langdon. *Aborigines, Land and Land Rights*. Canberra: Australian Institute of Aboriginal Studies, 1983.

Petroff, Ivan. *Population and Resources of Alaska*. House Executive Document 40, 4th Congress, 3d session, 1881.

———. *Report on Population, Industries and Resources of Alaska*. Washington, DC: U.S. Government Printing Office, 1882.

Picard, Pirmin. "Souvenirs de Voyages le Klondyke." *Le Monde Illustré* 14, no. 696 (September 4, 1897): 294, 297.

Québec Bibliothèque et Archives Nationales. Revues d'un Autre Siècle, Photograph No. 3053: "Frs. Mercier, Célèbre Voyageur Canadien." *L'opinion publique* 2, no. 43 (October 26, 1871): 517. http://bibnum2.banq.qc.ca/bna/illustrations/accuiel.

———. Revues d'un Autre Siècle, Photograph Nos. 1794, 1795, 1798, 1801: "Scènes du Klondyke: Station Mercier (Station Tanana)." *Le Monde Illustré* 14 no. 696 (September 4, 1897): 297. http://bibnum2.banq.qc.ca/bna/illustrations/accuiel.

Raboff, Adeline Peter. *Inuksuk Northern Koyukon, Gwich'in & Lower Tanana 1800–1901*. Fairbanks: Alaska Native Knowledge Network, 2001.

Raymond, Charles. *Report of a Reconnaissance of the Yukon River, Alaska Territory. July to September, 1869*. Washington, DC: U.S. Government Printing Office, 1871.

Reclus, Élisée. *Nouvelle Géographie Universelle La Terre et Les Hommes xv Amérique Boréale*. Paris: Librarie Hachette et Cie., 1890.

Ryden, Kent. *Mapping the Invisible Landscape: Folklore, Writing, and the Sense of Place*. Iowa City: University of Iowa Press, 1993.

Schmitter, Ferdinand. *Upper Yukon Native Customs and Folk-lore*. Washington, DC: Smithsonian Institution, 1910.

Schneider, William. "Chief Sesui and Lieutenant Herron: A Story of Who Controls the Bacon." In *An Alaska Anthology: Interpreting the Past*. Edited by Stephen W. Haycox and Mary Childers Mangusso. Seattle: University of Washington Press, 1996.

Schwatka, Frederick. *Along Alaska's Great River*. Anchorage: Alaska Northwest Publishing Company, 1983.

Scott, Elva R. *Jewel on the Yukon, Eagle City: Collection of Essays on Historic Eagle and Its People*. Eagle City, AK: Eagle Historical Society and Museums, 1997.

Sha Tan Tours. *Yukon First Nations Map: The Real Map According to Our Elders*. Whitehorse, YT: Sha Tan Tours, c. 1998.

Sherwood, Morgan. *Exploration of Alaska 1865–1900*. Fairbanks: University of Alaska Press, 1992.

Sidney, Angela, Kitty Smith, and Rachel Dawson, *My Stories Are My Wealth*. Whitehorse, YT: Council for Yukon Indians, 1977.

Sparks, W. F. *A Friend to Man: The Story of Frank Sparks*. Victoria, BC: W. F. Sparks, 2006.

Teben'kov, M. D. *Atlas of the Northwest Coast of America*. Translated and edited by R. A. Pierce. Kingston, ON: The Limestone Press, 1981.

Trepanier, Leon. "L'Aventureuse Carrière de Moïse Mercier, de St-Paul L'Ermite, le Premier de Race Blanche à Explorer les Rivières de l'Alaska." *La Patrie* (October 1, 1950): 38–9.

Turck, Thomas J., and Diane L. Turck. "Trading Posts along the Yukon River: Noochuloghoyet Trading Post in Historical Context." *Arctic* 45, no. 1 (March 1992): 51–61.

U.S. Coast and Geodetic Survey. *United States Coast Pilot 8: Pacific Coast. Alaska. Dixon Entrance to Cape Spencer*. Washington, DC: U.S. Government Printing Office, 1869.

U.S. Congress. *Senate Executive Document 143. Message from the President of the United States transmitting a Report of the Secretary of State Relative to the frontier between Alaska and British Columbia*. Readex Archive of America, U.S. Congressional Serial Set, 1817–1980.

———. Senate Committee on Military Affairs. *Senate Report 1023, 56th Congress, 1st session, 1900 Compilation of Narratives of Exploration in Alaska. Compilation of Explorations in Alaska*. Washington, DC: U.S. Government Printing Office, 1900.

U.S. National Park Service. *Yukon-Charley Rivers National Preserve*. Washington, DC: U.S. Government Printing Office, 2005.

Webb, Melody. *Yukon: The Last Frontier*. Vancouver: University of British Columbia Press, 1993.

Whymper, Frederick. *Travel and Adventure in the Territory of Alaska, Formerly Russian America—Now Ceded to the United States—and Various Other Parts of the North Pacific*. London: John Murray, 1869.

Wickersham, Honorable James. *Old Yukon Tales—Trails—and Trials*. Washington, DC: Washington Law Book Co., 1938.

Wright, Allen A. *Prelude to Bonanza*. Whitehorse, YT: Studio North, 1992.

Yukon Archives. "Yukon with a French Touch." Yukon Archives. http://www.tc.gov.yk.ca/archives/frenchyukon/voyag/voyag2.html (accessed March 6, 2007).
Yukon Historical and Museums Association. *Discovering Northern Gold Symposium Program*. Yukon Historical and Museums Association, Whitehorse, YT, October 2006. http://www.yukonalaska.com/yhma/ngold.pdf (accessed March 6, 2007).
———. *The Kohklux Map*. Whitehorse, YT: Yukon Historical and Museums Association, 1995.
Yukon Native Brotherhood. *Together Today for Our Children Tomorrow*. Whitehorse, YT: Yukon Native Brotherhood, 1973.
Yukon Native Language Centre. *Hän Literacy Session*. Whitehorse, YT: Yukon Native Language Centre, 2006.

Unpublished and Archival Sources

Canada. Geological Survey of Canada. Manuscript Map H2/600/1888 NMC023871, Map Collection, Library & Archives Canada, Ottawa, ON.
Canada. Statistics Canada. Census taken at Rampart House, 1891. MF68, reel no. 1–2. Census, Yukon Territory, 1901. MF81, reel no. 1–3. Yukon Archives, Whitehorse, YT.
Cooke, C. "A New and Accurate Map of North America Including Nootka Sound; with the New Discovered Islands on the Northeast Coast of Asia." Rare Map G3300 [1780] C66, Alaska & Polar Regions Collections, and in Online Collections, "Meeting of Frontiers" rare maps Web site of the Alaska & Polar Regions Collections, Rasmuson Library, University of Alaska Fairbanks.
Dall, William Healey. "Alaska and Yukon Territory." Rare Map UAF G4370 1869 U55, Alaska & Polar Regions Collections, Rasmuson Library, University of Alaska Fairbanks.
Davidson, George. Correspondence, diaries, maps, photograph albums. The Bancroft Library, University of California, Berkeley.
Great Britain. "British War Office Map of Alaska and Adjoining Regions." Map Collection H3/1209/Alaska/1886. Library and Archives Canada, Ottawa, ON.
Isaac, Gerald. "The Han Huch'inn Early Warning System." Manuscript in personal collection of Linda Johnson.
Jesuit Mission Records, MF no. 96, roll 21, Frame 58, Nulato, 1875. Alaska & Polar Regions Collections, Rasmuson Library, University of Alaska Fairbanks.
Juneby, Willie. "Place Names of the Eagle Region." Transcribed by John Ritter, August 1978. Yukon Native Language Centre, Whitehorse, YT.
Kandik, Paul, and François Mercier. "Map of Upper Yukon, Tananah and Kuskokwim Rivers." Map Collection G4370 1880 K3 Case XB, The Bancroft Library, University of California, Berkeley.
Karta Rossiiskikh Baabhih. Rare Map G4370 1861 K37. Alaska & Polar Regions Collections, Rasmuson Library, University of Alaska Fairbanks.

Kingsbury, Willis V. "Yukon River Diary," Vols. 1 and 2, 1889–91, transcribed by Jim Paull, February 2008. Copy in personal collection of the author.

Kokrine, Effie. Tape no. H2001-104, October 25, 2001. University of Alaska Archives, Oral History Collection, Alaska & Polar Regions Collections, Rasmuson Library, University of Alaska Fairbanks.

McDonald, Kenneth. Journals and correspondence, 1875. Letters and Papers of the Church Missionary Society, London. Typescript copy, Yukon Native Language Centre, Whitehorse, YT.

McDonald, Robert. Journals and correspondence, 1862–1913. Letters and Papers of the Church Missionary Society, London. Typescript copy, Yukon Native Language Centre, Whitehorse, YT.

———. "Ven R. McDonald," manuscript life story transcribed by Rev. T. G. A. Wright, 1911. Robert McDonald Fonds, Yukon Archives, Whitehorse, YT.

McDougall, James. Sketch Map of Yukon River. Hudson's Bay Company Archives, Manitoba Archives, C38/25 fo. 54 (N11325), Winnipeg, MB.

Moore, Patrick. "Archdeacon Robert McDonald and Tukudh (Gwich'in) Literacy." Manuscript under review for publication in *Anthropological Linguistics*, March 2008. Copy in personal collection of Linda Johnson.

Petroff, Ivan. Correspondence. The Bancroft Library, University of California, Berkeley, California.

Raymond, Charles. "The Yukon River, Alaska from Fort Yukon to the Sea." Rare Map digital ID f87101 http://hdl.loc.gov/loc.ndlpcoop/mtfxmp.f87101, Online Collections, "Meeting of Frontiers" rare maps Web site of the Alaska & Polar Regions Collections, Rasmuson Library, University of Alaska Fairbanks.

Sim, Vincent. Journals and correspondence, 1882–1885. Letters and Papers of the Church Missionary Society, London. Typescript copy, Yukon Native Language Centre, Whitehorse, YT.

United States. Twelfth Census of the United States, 1900, Alaska, MF rolls 17 and 18. Alaska & Polar Regions Collections, Rasmuson Library, University of Alaska Fairbanks.

Index

Italicized numbers indicate images.

A

B

C

G

H

I

J

K

L

M

N

O

P

R

S

T

U

W